Eduardo Vatuva

Development of the "SuperMente" Mobile Application

Eduardo Vatuva

Development of the "SuperMente" Mobile Application

Development of a Mobile Application for Cognitive Behavioral Psychotherapy Care

ScienciaScripts

Imprint

Cover image: www.ingimage.com

This book is a translation from the original published under ISBN 978-620-6-75761-0.

Publisher:
Sciencia Scripts
is a trademark of
Dodo Books Indian Ocean Ltd. and OmniScriptum S.R.L publishing group

120 High Road, East Finchley, London, N2 9ED, United Kingdom
Str. Armeneasca 28/1, office 1, Chisinau MD-2012, Republic of Moldova, Europe
Printed at: see last page
ISBN: 978-620-8-29619-3

Contents

DEDICATION

I dedicate this work to my parents **Alberto Mario Vasco Miguel Vatuva** (in memory) **and Cacilda Barbosa Sachiambo**, for having always created the conditions to support my academic training. In difficult times, they didn't hesitate, spared no effort and always did their best, depending on their human and financial capacities.

To my wife **Joyceline Ofelia Essanju Kassule Chiyaka Vatuva**, who always supported and believed in me when I made the decision to study for a master's degree.

My family, friends and colleagues who directly or indirectly made this work possible.

ACKNOWLEDGEMENTS

I thank God, the source of life, for giving me health, protection and wisdom during my academic career.

To my parents, siblings, especially **Alzira Baca Sachiambo Vatuva** and **Ilidio Paulo Sachiambo Vatuva**, who have shown unconditional support from the very beginning; for long hours we have spent reflecting on certain subjects related to the work. To my colleague Bernardo Kudikudilema who spared no effort in sharing his knowledge for the conception of the project.

My wife **Joyceline Ofelia Essanju Kassule Chiyaka Vatuva**, who was always available to help.

To the psychology professionals who always responded positively to the questions that were asked, which allowed the work to be carried out.

To the Angolan State for providing a scholarship through INAGBE.

My colleagues, in general, did their best to share their knowledge throughout the course.

To Professors MSc. **Angel Correa** and MSc. **Yosbel Camacho Izquierdo** for their instruction and guidance in developing the end-of-course work. Thanks also go to Professors: Dr **Mbona Samuel Antonio**, MSc **Irina Diaz Torres**, MSc **Oleiny Carrasco**, MSc **Karina Kollazo Oliva**, MSc **Grisel Pino**, Dr **Wilfredo Falcon Urquiaga** and others.

The Gregorio Semedo University, for giving me the opportunity to study for a Master's degree in the speciality of Mobile Application Development.

To everyone who directly or indirectly contributed to the completion of this work.

EPIGRAFE

"Studying is not an act of consuming ideas, but of creating and recreating them."
Paulo Freire

SUMMARY

The development of technology has made a positive contribution to the advancement of science and technology. Psychology has taken advantage of information and communication technologies, as it uses technological resources that serve as an alternative means of distance communication. Angola has only tentatively shown some progress in the use of new information and communication technologies.

The level of user maturity has grown considerably, which is a plus, as it allows developers to improve their techniques in order to respond accurately to the demands of the market. The demand for online services has grown considerably, and mental health professionals are finding an opportunity to assert themselves in the market. This open environment allows professionals to address their problems without taboos. The aim of this work is to develop a mobile application for cognitive behavioural psychotherapy. Despite being new to Angolan society, the service of online cognitive behavioural psychotherapy is well known in developed countries. The implementation of the mobile application arose from the need for a fringe of Angolan society to feel excluded because they didn't know how to solve their problems without being present in the counselling room. There are other reasons why this project was conceived, such as: demand for services, scheduling incompatibility, inefficient management of follow-ups, lack of communication between professional and patient.

Keywords: Mobile application, service, cognitive and behavioural.

CHAPTER I

1 INTRODUCTION

The emergence of mobile applications has revolutionised information and communication technologies and significantly altered the life of contemporary man, bringing him closer to the globalised world. This technological innovation has not come about by chance, but as a result of in-depth thinking based on the use of mobile devices such as smartphones and tablets.

Developing a mobile application for online care in cognitive behavioural psychotherapy is a topical issue, as the most developed countries continue to improve their technological solutions in order to comply with the legal system. Existing solutions have not been standardised; each country has developed them according to its local reality.

Professionals will certainly be able to make better use of technological resources and find other ways of working in order to improve their performance at work and make it more dynamic, for example: shortening distances, reducing costs, improving communication, increasing production.

The elements mentioned above have served as a barrier to improving the provision of services in cognitive behavioural psychotherapy, as well as the professional growth of psychotherapists. The emergence of this tool could translate into an alternative means of communication for various situations that are still present in mental health institutes.

This technological solution can serve as an alternative source for thousands of people who still suffer in silence because they can't find the best solutions.

Problematic situation

Psychological care for patients suffering from mental disorders is still a very bureaucratic process; the administrative procedures used in private and public institutions are time-consuming. The institutes that provide services in cognitive behavioural psychotherapy have provided an inefficient service, and it is clear that they use formal procedures aided by traditional methods, which has made life difficult for patients.The Angolan population, who live in urban areas, have little information about mental health, despite the existence of the Internet on smartphones, tablets, laptops and desktops. There are also few professionals who organise meetings and, in turn, pass on information in psychological forums, and there is a lack of interaction between professionals and patients outside the clinic.

The search for psychological services seems to be mostly motivated by suffering, a moment of pain, anguish, doubt and often hope. In Angolan society, there are high rates of patients who resist contacting mental health professionals. In some cases, the patient is excluded by society through discriminatory practices, which can cause deep crises in their emotional life, such as: reduced self-esteem, personality disorders, anxiety, depression and feelings of guilt. Traditional psychotherapy has not helped with the aspects mentioned because it is too closed off; it is important to create other approaches to improve mental health services. Therefore, there is a need to deconstruct the concept that going to the psychologist implies being crazy. To this end, it will be up to psychology professionals to review other forms of action in order to regularly inform society about the importance of mental health.

Scientific problem

How can we improve the service in cognitive behavioural psychotherapy?

Object of study

The cognitive behavioural psychotherapy service.

Field of study

The study to develop the mobile application for the online service of cognitive psychotherapy was carried out in the Province of Luanda, Talatona Municipality, Benfica Neighbourhood, Rua da Clemencia, House No. 17, at the CRER E SER Clinic **(See Annex XIX).**

Objectives

General objective

Developing a mobile application for online cognitive behavioural psychotherapy.

Specific objectives

1. Structuring the data obtained to support the development of a mobile application for the online service of cognitive behavioural psychotherapy;
2. Documenting applications based on software engineering techniques;
3. Implement the functionalities of the mobile application;
4. Testing the SuperMente mobile application, version 1.0

Hypothesis formulation

The mobile application for online care in cognitive behavioural psychotherapy guarantees improvements in information and communication in real time between the patient and the psychology professional. The inclusion of information and communication is a way of solving the problem in the provision of cognitive behavioural psychotherapy services.

H1- Based on the scientific problem, the following hypothesis was created: if the "SuperMente" mobile application is developed, it will improve the quality of care in cognitive behavioural psychotherapy.

Methodologies and procedures

Types of research used

Research objectives

Exploratory research: allows you to familiarise yourself with the problem. This research includes: a bibliographical survey, interviews with experienced people. Normally, the research is bibliographical and a case study (Gil, 2008).

Descriptive research: describes the characteristics of phenomena or populations. One of its peculiarities is the use of defined data collection techniques, such as questionnaires and systematic observation (Gil, 2008).

Research into technical procedures

Bibliographical research: based on the material prepared, mainly scientific articles and books (Gil, 2008).

In the initial phase there was a need to use the descriptive method, given its limitations, bibliographical research was used and extensive direct observation and interviews were used.

In the second phase, the systematic method was used to enable the study of objects; each part of the object and its component to be studied in detail and the relationship between them obeyed.

To finalise the agile development methodology, an overview of XP (Extreme Programming) will be given, allowing a broader approach to the development of agile methodology. The XP process uses an object-orientated paradigm and involves a set of permanent rules and practices in the context of four methodological activities: planning, design, coding and testing. An in-depth approach has been taken to software development methodology **(see Chapter III).**

Population and sample

This work is the result of information sharing between mental health institutes (psychology clinic and mental health centre), as well as psychology professionals who responded on their own behalf, but it is an unfinished product. The survey was aimed at psychology professionals, and the variables used were: age group, gender, academic level and occupation.

During the data collection phase, a total of 54 of the population interviewed was obtained, which corresponds to 100%.For the gender variable: 61.1% represents the female population while 38.9% is male; age group: 13% represents the younger population, whose ages range from 24 to 30 years of age; 44.4% of the population, represents ages ranging from 31 to 40 years of age; 40.7% of the population represents ages ranging from 41 to 60 years of age; occupation 31.5% represents the worker-student population while 59.3% are workers; as for the academic degree: 7.4% represents the population with a bachelor's degree, 66.7% a bachelor's degree, 20.4% a master's degree and 5.5% a doctorate; level of acceptance: 75.9% of the population answered yes while 22.2% maybe and the rest answered no, as illustrated in the graphs below, **(See Annex I)**.

Project structure

The work is divided into five chapters: introduction, problematic situation, scientific problem, object of study, field of study, general objective and specific objectives, formulation of hypotheses, methodologies and procedures, population and sample.In Chapter II, a bibliographical survey was carried out of scientific works produced by teaching institutes and by well-known professionals in the field, which served as a material basis for the theoretical foundation and to put the practical component into perspective by reviewing the literature. Chapter III deals with aspects related to development methodologies, the tools used, requirements, diagrams and system architecture. Chapter IV shows the results of the project and the tests carried out. Chapter V presents the conclusions and recommendations. The elements that make up the last part are the bibliographical references, appendices and annexes.

CHAPTER II

2.LITERATURE REVIEW

2.1 Concept of psychology

The word psychology is a nomenclature that originated from two Greek words: psyche, which refers to the mind, soul or spirit, and logos, which refers to the study of a particular subject. Psychology is the science that studies the human mind through behavioural techniques. From the beginning of the 18th century, the term psychology began to be coined, recognised and came to mean "the study of the mind". In contrast, psychology emerged as a scientific discipline at the end of the 19th century (Weiten 2011, quoted by Leite, 2016, p. 15).

Psychology studies human and animal behaviour, using specific techniques to achieve three fundamental objectives: the description, prediction and control of behaviour (Teles 1995, cited by Correa, 2013, page 16).

2.2 Psychological assessment

Since the dawn of vocational psychology, psychological assessment has been a tool to help individuals make decisions about their lives (Duarte, 2008, quoted by Correa, 2013, p.16).

Psychological assessment is a process that aims to awaken the professional to devise strategies, create material and human conditions incorporated into the process to solve a particular problem presented by the patient. The assessment process begins before the patient presents their mental health problem (Tavares, 2012, quoted by Correa, 2013, page 16).

2.3 The history of online customer service

Online psychotherapy activity is believed to have begun in 1960, using software called Eliza, created by Joseph Weizenbaum at ITM. The programme's implementation simulated a conversation between a patient and their psychologist, where the user was the patient and the software was the psychologist (Lemos, 2016, cited by Munhoz, et al, 2019 page 2).

The service can be carried out in real time, for example: video conference conversations, instant messaging via chat and email exchanges. Video conferencing is the closest thing to face-to-face contact when it comes to quality and results. This procedure makes it possible to gather as much sensory information as possible, such as hearing and sight. In addition to the aforementioned elements, even if there is a screen between the psychotherapist and the patient, the video conference allows both angles to benefit in terms of anxieties, feelings and emotions (Lemos, 2016, quoted by Munhoz, et al, 2019 page 2).

2.4 Man and technology

The emergence of post-modern societies began in the 1950s and has been characterised in human history as the cultural industry has created new products: information technology, the Internet and virtual reality. These inventions have produced social development, economic growth and improved human relationships (Mrech, 2000, cited by Goncalves; Belmino, 2017, page 33).

The Internet marks a new era in the field of interpersonal interaction, making it acceptable for unknown people to get to know each other in a short space of time in a virtual environment, as well as getting to know the world virtually: countries, cultures, museums, libraries, among others. It was against this backdrop that it was possible to create the first chat rooms (Webchats, IRC), which became a major attraction, together with an inexhaustible source of knowledge, and were responsible for the intense spread of the Internet. The expansion of this new communication paradigm has led to massive access to browsing and interactive programmes, creating major changes in the modus operandi, functioning and organisation of subjectivity in the interplay between the real and virtual planes. In this context, the computer plays an insightful role in data processing, targeting and accessibility of information (Almeida; Eugenio, 2006, cited by Goncalves; Belmino, 2017, page 34).

History tells us that the estrangement between man and technology began in ancient Greece, when Plato differentiated between what is produced by man (tekhne) and what is produced by nature (phusus). After differentiating the tekhne from the episteme, theoretical thought, only available to philosophers, all those who worked from the tekhne were labelled as inferior. However, it was the industrial revolution that influenced the emergence of steam engines and the myth that they could dominate the world, widening the gulf between man and technology (Belmino 2010, cited by Goncalves; Belmino, 2017, page 34).

Technology emerged in human history as an imperative necessity, but it was with the industrial revolution that profound and significant changes in social dynamics took place, where the creation of a new urban space with the formation of metropolises transformed the way people lived (Costa 2002, cited by Goncalves; Belmino, 2017, page 34).

In the technological revolution, we realise that the changes in social relations, where the Internet is made available as a tool for human interaction in cyberspace, have created new and unknown life experiences (Leitao; Costa, 2003, cited by Goncalves; Belmino, 2017, page 34).

2.5 Prospects and challenges

The emergence of the new paradigm makes it possible to communicate in contemporary society, as well as creating online access mechanisms that serve as an alternative for expanding the psychotherapy service. Faced with these innovations, professionals are faced with some concerns, such as: receiving and sending messages by e-mail and instant messaging via Internet communication software, among other questions that serve as a basis for the psychologist's clinical practice in a new, more interactive approach in the psychology counselling room. (Barbosa et al, 2013, cited by Ulkovski et, al,2017, page 60).

Online psychotherapy offers some conditions for it to be carried out properly. One of them works as follows: the patient and the professional must be familiar with the technology to be used, otherwise online counselling will become an unfeasible process for its intended purpose. Online care itself can take longer than face-to-face care. The trade-off is the patient's and professional's freedom of choice and expression (Costa, et, al 2009, quoted by Ulkovski, et al,2017, page 62).

Online psychotherapy presents a number of problems, most notably voice intonation, understanding silences and pauses, problems connecting to the internet, and a lack of knowledge about how to use technological tools (Pinto, 2002, quoted by Ulkovski,

et al, 2017, p.62).
In some cases, the professionals manage to achieve the objectives proposed in the activity, however, the online guidance is only the gateway, the possibility of diagnosing and referring the patient for effective care, which can be a significant change in the patient's life, since it is considered as a technical procedure, the new modality of care. (Fortim; Consentino,2007, cited by Ulkovski, et al,2017, pag.62).
There is a demand for psychological services. Research shows that there are no different rates for face-to-face and online counselling. Online psychotherapy is gaining ground as time goes by, and there are reasons for this, such as the convenience, efficiency and security that technological tools offer. Cognitive behavioural psychological treatment for various disorders, such as panic disorder, agoraphobia, post-traumatic stress disorder and depression, has created some discomfort when done in person. Professionals' experience of face-to-face counselling shows that patients are nervous at the start of the consultation, which inhibits them from providing accurate information to solve the problem. This phenomenon, as well as being a concern for psychology professionals in terms of improving their technical and deontological skills, is a sign for information and communication technology professionals who must constantly study and create ideas to design solutions capable of meeting the challenges. (Pieta; Gomes 2014, cited by Lins, 2020, page 144).

2.6 Limitations and possibilities as perceived by online psychotherapists

At the start of the 21st century, the influence of technology on people's daily lives has been recognised. The strong presence of technological tools in people's lives has brought the world closer together, increased fluidity in communication, created an environment of proximity between people living in different areas, distanced people who share the same space, changed people's psychological behaviour, created an open and limitless space for the dissemination of information, improved alternative sources of communication. Rapid technological progress has led to an adaptation of science in search of new knowledge. The discovery of the Internet has intensified the debate on the study of psychology in its essence by making use of experience and technology, which has led to the topic attracting the attention of the academic and professional community (Marot; Ferreira, 2008, cited by Pinhatti, 2013, page 7).

2.6.1 Internet psychotherapy studies

Studies have shown that online therapies have a very high level of effectiveness and efficiency when compared to traditional therapy. Patients consulted in traditional therapy show some resistance in terms of interaction with the professional. The professional needs more information to make decisions. In turn, the patient is reluctant to answer certain intimate questions, without taking into account the specific characteristics of the illness. Therapy carried out over the internet has many benefits for the patient and can produce better results (Shapira, et al, 2008, quoted by Pinhatti, 2013, p.10).
Technology has energised societies to create applicable technologies by stimulating knowledge and increasing skills in the field of clinical psychology. Clinical psychology has also opened up the possibility for professionals to be evaluated by their patients using these tools. Information technology in psychological therapy divides the subject

into four essential areas:

1) Clinical development enables greater efficiency in data collection and processing;
2) The use of computer tools makes it possible to simulate the controlled environment, virtual reality and the acquisition of skills and abilities;
3) Online therapy has emerged as a new approach to solving interpersonal problems in the psychological forum, and not just for help.
4) The continuous training of mental health professionals involves the use and production of multimedia teaching materials, as well as the use of technological tools in general (Penalosa 2009, quoted by Pinhatti, 2013, page 8).

Online psychotherapy is divided into two parts: synchronous and asynchronous. Synchronous works on the basis of real-time communication, i.e. it follows the scheduling of a limited appointment at a counsellor's office. Asynchronous makes it possible to realise a communication that works on the basis of non-simultaneity, at a contractually defined time, within 24 hours. (Marot; Ferreira, 2008; Pinhatti, 2013, p.9).

2.6.2 Perspectives on online psychotherapeutic care

Online psychotherapy is based on technological tools, namely: instant messaging, audio, video conferencing, internet communication software, chat, camera and email. The use of these technological tools does not preclude the complement of face-to-face psychotherapy (Macdonald, 2007, quoted by Ulkovski, et al, 2017, p.61).

Online therapy creates mechanisms that are easy to access for people who, for various reasons, are unable to leave their home, whether due to difficulties with mobility or health, particularly in the case of the elderly or people with physical disabilities, factors that make it impossible to continue attending in person (Pinto 2002, quoted by Ulkovski, et al,2017, p.61).

The online service itself can take longer than the face-to-face service. However, this time is somewhat compensated for, as the testimonies do not need to be written down, and can be saved in Integra, bringing with them reliability in writing without fear of misinterpretation. (Costa, et al, 2009, cited by Ulkovski, et al,2017, pag.62).

2.6.3 Online psychotherapy approaches

The approaches most commonly used in online care are those based on psychoanalysis and cognitive behavioural therapy. Although it is not precise how these approaches work in the virtual environment, it is possible to identify the effectiveness of online CBT, which has demonstrated greater suitability in the treatment of illnesses such as: depression, anxiety, panic disorder, phobia and post-traumatic stress (Anderson, et al 2002, cited by Ulkovski, et al,2017, pag.62).

Psychological professionals can provide online counselling using technological tools. Counselling differs from psychotherapy in that it is a one-off service with the aim of providing the user with precise information and, if necessary, referring them to other services. Counselling is part of the therapeutic bond, a common factor in any psychotherapeutic approach (Bock, 2000, quoted by Hermosilla, 2008, p. 2).

Several questions have been raised in relation to these problems, such as:

Familiarity with the use of communication technologies: the professional carrying out this activity must be **familiar** with communication tools and cutting-edge technologies;

Specific training: professionals must be able to receive specific training for online

service;
Security and privacy: with regard to the security of communications, care must be taken at all times to use encryption when transmitting data, as well as to inform the patient of the need to be in a private, uninterrupted physical location before starting care. It is up to the professional to ensure that during data collection, the information is kept confidential, and the professional is the sole custodian of the information. If the information ends up in the hands of a third party, the patient has the right to take legal action and bring a criminal case against the professional concerned;
Identifying the patient and the professional: another question that arises is how to guarantee the identification of the parties involved in the service. How do you know if a professional is really on the other side? How can we guarantee that the patient really has the identity reported? To deal with these issues, digital authentication systems must be used and possibly registration information sent to the psychologist that is not normally required in a face-to-face consultation;
Handling crisis situations: how should professionals act in the event of a patient crisis if they are in a remote location? For this problem, it is important that the professional has alternative contact details for their patient so that in the event of a crisis, they can contact their relatives so that they can intervene as soon as possible.
Connection problems: Internet connection drops occur frequently, as do software and operating systems which, from time to time, crash and malfunction for no apparent reason. These problems could interfere with the service if it is being provided via chat or video conference. Video conferencing services can present connection difficulties, especially when the broadband speed is at a certain level. The use of asynchronous technologies, such as e-mail, practically eliminates the problem of interruptions in the service and the adoption of broadband access may allow better use of video conferencing for psychological counselling.
Lack of non-verbal stimuli: communication via text limits the professional from making an in-depth diagnosis; In this case, the professional faces the problem of not having any data about the patient at their fingertips, apart from their writings, such as the patient's image, body posture, facial expression, tone of voice, clothing, among others. This lack of data can often increase the chances of interferences in communication and this can lead to a lack of understanding in the exchange of information, thus causing, among other things, diagnostic problems.

2.6.4 Online psychotherapy and the therapeutic relationship

The lack of mastery in handling technological tools has led to a lack of knowledge of computerised solutions, which can be considered reluctance. Professionals who believe that it is impossible to carry out a therapeutic consultation outside of a face-to-face meeting are considered to be reluctant, making them believe that the virtual meeting detracts from the quality of the results. There is a group of professionals who believe that it is impossible to parameterise diagnoses using digitised words based on emotions and feelings. The others claim that the phenomena can only be transferred from the patient to the professional through a face-to-face consultation and not a virtual one. Some obstacles have been taken into account regarding the establishment of a bond in virtual consultations, such as: distrust of the services provided on the internet, lack of mastery of the technologies used to provide care or a preference for traditional consultations (Pinto 2002, quoted by Ulkovski, et al,2017,

p.63).
The therapeutic relationship used in psychotherapeutic interventions shows that online therapy does not differ significantly from traditional therapy.
Some therapists report that there is a high level of trust and have identified it quickly compared to face-to-face consultations. Patients are more open about the difficulties they are facing, have a more uninhibited attitude and feel a sense of fulfilment compared to conventional consultations (Pieta; Gomes, 2014, quoted by Ulkovski, et al, 2017, p.63).
In the traditional clinic, psychologists are trained to listen, but in virtual consultations, they are given a new way of listening and interpreting facial messages, because the way the message is written and the words are used makes it possible to observe feelings and emotions. The bond established in a virtual consultation can exist regardless of whether it is online or face-to-face. During online psychotherapy, the relationship can become a resource for note-taking, given the amount of information transmitted by the patient. It is necessary to pay special attention to what is said and written, compared to what is observed, since this observation is not always accurate (Siegmund; Lisboa 2015, cited by Ulkovski, et al,2017, page 64).
Virtual counselling is not intended to be a new form of psychotherapy, but a new approach to psychotherapeutic intervention. This intervention consists of solving a problem in a different psychological forum to the one the patient is normally in, making use of two-way communication, which can be carried out by text, audio or audiovisual means. Examples: chat, videoconferencing and e-mail (Kaplan 1997, quoted by Pinhatti, 2013, p.9).

2.6.5 Ethics in the context of online psychotherapy, questions and reflections

The psychologist must be able to receive the foreigner, in their peculiarities, and take on the risks of the process. The subject in the therapeutic process bears the marks of a world full of rapid changes in values, so that they are moulded by current changes. This is a challenge for psychotherapists, the fact that they don't find the limitations to establish an efficient and satisfactory clinical practice in the face of the phenomena observed. (Magdaleno 2010, quoted by Ulkovski, et al,2017, pag.64).
There is an urgent need in the treatment of some psychotherapeutic disorders; the use of a more flexible attentive eye in the virtual consultation, if this type of treatment does not involve alterations in the patient's psycho-emotional state, can generate a quick response, reduce anxiety and give the possibility of a face-to-face meeting. (Pinto 2002, quoted by Ulkovski, et al,2017, pag.64).
Online psychotherapy makes communication more effective, which facilitates intervention in the behaviour of schizophrenia, anorexia, rape episodes, incest, guidance for solving problems such as excessive use of alcoholic beverages that can lead to suicidal behaviour (Goldberg 1992, cited by Ulkovski, et al,2017, page 64).
Reflecting on the authorisation of online psychotherapy, does taking responsibility for the whole process and the risks that come with it equate to face-to-face therapy? Isn't creating restrictions on this practice making it more difficult for people who live in remote areas where there are no professionals or who don't have lifts to get to the counsellor's office? What would be contrary to ethics is to make it impossible for therapists to carry out virtual consultations due to a lack of scientific knowledge for people who need psychological intervention? (Suler, 2002, quoted by Ulkovski, et

al,2017, p.64-65).
One of the challenges that psychology professionals may face, and the impact suffered by the professionals themselves, is prejudice, doubt and disorientation in relation to the effectiveness and technique of understanding their patients. The positive value of the media in disseminating mass information and creating contacts cannot be ruled out, but it is possible to analyse the alternative ways of understanding and dealing with the information. It is important to recognise that in the process there is a different subject from the psychic configuration. Psychology has a duty to seek understanding (Costa, 2005, cited by Ulkovski, et al,2017, p.65).
Psychotherapy is a meeting between two people, where one is willing to help through science and technique, professional help allows emotions and feelings to be exposed. This complex dynamic ultimately affects both of us, as it creates a space for profound change in the person's life. Online psychotherapy can trigger a disorientated environment when both the professional and the patient are overwhelmed by emotions. This also happens when reading a book, where fantasy takes over the real world, as the reader is taken into a world based on the author's imagination (Pinto 2002, quoted by Ulkovski, et al,2017, p.65).

2.6.6 Possibilities and risks of online counselling

The rapid changes caused by the frequent transformations in the technological world require constant reflection in order to keep up with contemporary changes. In the field of social relations, we can observe the emergence of a virtual environment made by man, through his relationship and appropriation with the new technological apparatus, where interaction takes place even if the parties are far from each other. (Fortim; Cosentino 2007, cited by Goncalves, et al, 2017, page 36).
Networked communication is invading people's daily lives in post-modern times; the impact of this reality extends to psychology, which finds a field of study and action in this environment. Faced with these changes in this space, adaptations are also needed to meet this growing demand for online psychology services, along with a growing range of services (Farah, 2004, cited by Goncalves, et al, 2017 page 36).
The modalities of virtual consultation offer a path that began with the computerisation of psychological tests where real therapist assistants were projected, followed by the specialised *homepage* containing information on specialised topics such as career guidance and psychopathology. Then came a third type of psychological intervention on the Internet, the virtual psychological consultation. Limited in number, with a neutral and informative character, it has a higher level of acceptance than face-to-face consultations. The initial contacts are prolonged, through ongoing consultations (Pinto 2002, cited by Goncalves, et al, 2017, page 36).
The research produced positive considerations regarding the establishment of the therapeutic alliance in the realisation of psychotherapy through asynchronous messages. Most of the time, communication via the Internet takes place in
real time. The inability to observe the non-verbal discourse is seen as very significant in the decoding of the report, especially when the service is carried out via e-mail, where the patient can elaborate and re-elaborate the text before sending it to the therapist (Prado; Meyer 2006, cited by Goncalves, et al, 2017, page 36).
The psychologist has an obligation to inform the patient about the technological resources used to guarantee the confidentiality of information and to clarify the

possible risks, since electronic correspondence is subject to invasion. These measures must be taken in order to protect it, such as encrypting messages, the coding process and privacy in the physical location where the communication instrument is kept (Prado 2000 cited by Goncalves, et al, 2017, page 37).
The Internet has promoted the democratisation of information, and can favour access to psychology services for people who previously had no access to these services, such as the elderly, the physically disabled, professionals who travel constantly or people who are isolated and in remote regions. The virtual modality of psychological practice is not a simple transcription of traditional practice into virtual space, but an expansion of the psychologist's work that is directly related to the historical moment in time. This practice makes it possible to create the conditions to expand the audience for their interventions, since access to the Internet has become increasingly widespread. It is essential that both the patient and the psychologist feel comfortable handling the technological tools in order to take advantage of them. Psychological professionals should have constant training in the use of technological tools in order to maintain the same standard of quality in online care, which will ensure greater adherence from their clients compared to face-to-face care. (Fortim; Consetino 2007, cited by Goncalves, et al, 2017, page 37).

2.6.7 Concerns about the online psychotherapy service

In virtual counselling, there are still some reservations and resistance on the part of professionals, with regard to cultural problems (regionalisms), problems related to licencing and jurisdictional laws in the modality of psychological care compared to face-to-face counselling. In addition to some doubts about confidentiality during the conversation between the professional and the patient, it is necessary to ensure the risks and benefits for the patient and the professional, among other ethical aspects involved (Finn; Barack, 2010; cited by Rodrigues, 2016, p.738).
There are developmental mechanisms in psychotherapy, such as relational work. Relational work is a mechanism for change and of great importance for patients (Hill; Knox, 2009; cited by Rodrigues, 2016, p.738).
Crisis situations involve risks and must be taken into account when indicating treatment and planning therapy (Tavares 2012, cited by Rodrigues, 2016, p.738),
As for interventions carried out through technological means, there is empirical evidence to suggest that people with stigmatising problems take pleasure in revealing intimate content, as is the case with drug users and people with sexual problems (Meier, 1988; cited by Rodrigues, 2016, p.739).
Asynchronous therapies should be aimed at people with psychotic disorders, people with difficulties in processing written communication. In the same way as patients with high levels of anxiety and distress, they require special attention when using online psychotherapy services, given the complexity of their condition. For these patients, synchronous counselling is a more appropriate form of professional help than asynchronous counselling, where waiting for a response can increase suffering or fail to relieve it in time.
Despite the distance, patients in a state of emotional distress can be safely approached and emotionally touched via synchronous online communication devices. Patients in crisis situations tend to share their experiences and feelings with anonymous people on the internet as a mechanism to reduce personal inhibitions.

The crisis intervention service, carried out by qualified psychotherapists, can be important in helping people who are in a state of distress and who feel drawn to the virtual space where they share painful experiences (Finfgeld, 1999; cited by Rodrigues, 2016, p.739).

2.6.8 Grounds in favour of virtual consultation

The analysis of the participants' comments focused on disinhibition and the feeling of freedom to express themselves in the virtual consultation without creating embarrassment or fear of the psychotherapist's judgement. Patients seen in person, on the other hand, expressed feelings of stress at the face-to-face service. Analysing the data from this study, 75% of the 15 patients seen in virtual consultations demonstrated skills, honesty and openness with the professionals. Some participants believe that online psychotherapy is more conducive to therapeutic changes compared to face-to-face consultations, due to the fact that physical distance facilitates disinhibition (Cook; Doyle 2002; cited by Rodrigues, 2016, p.739).
The anonymity that predominates in asynchronous online consultations makes it easier for people who have difficulties communicating in person to seek professional help, such as shy and insecure people. The physical distance between the professional and the patient is retained in synchronous communication, in the presence of a camera, which can reduce the discomfort of the shy patient or the patient who has a complaint about an embarrassing situation to be shared in person. People who value non-verbal communication make use of the microphone and camera, which allows the psychotherapist to maintain eye contact and perceive signs of body language, facial expression and tone of voice, in the same way as in the face-to-face modality (Esparcia, 2002, cited by Rodrigues, 2016, p.739).

2.6.9 Ethical aspects of online psychotherapy

The assessment of harm to treatment must take into account the benefits of virtual consultation. The existence of a risk should not necessarily inhibit the adoption of a technical or administrative procedure. Professionals have an ethical responsibility to provide accurate information and the services requested. There are still reservations about virtual consultation, as pressure from users who want the service at a distance can lead professionals to feel obliged to provide on-demand services. Otherwise, professionals will not fulfil their true role and patients will be forced to seek out people who are not qualified or licensed to do so (Childress 2000; cited by Rodrigues, 2016, p.740).

2.6.10 Benefits of online psychological services

Individuals living in different regions, most notably in large urban centres, the rural population, marginalised individuals, the prison population, people with physical disabilities, senior citizens, members of the same family who need family therapy. These can be some of the profiles of individuals who will benefit. As well as shortening the distance, it gives people the chance to choose the various therapeutic modalities, location and communication with the psychotherapist of their choice (Siqueira, 2014, quoted by Boscariol, 2020).

2.7 Psychological care in today's world

COVID-19 has had a major impact on people's mental health. Societies have adapted to the new reality, which shows the importance of implementing strategic services. Cost reduction, isolation, high levels of stress and suffering from insecurity have all been part of the lives of people who have tried their best to prevent the pandemic that is ravaging the entire world. The emergence of distance services has made technological tools a cross-cutting solution. Psychology was no exception to the rule, and part of the services that played its part in interventions used ICTs. This has awakened societies to value human capital in order to enable the marriage between man and machine. This has contributed to the reshaping of the mental health care landscape (FioCruz,2020, quoted by Boscariol,2020, page 8).

Online care is related to education, public health, publicising services and health care. Telepsychology is a segment of telehealth that uses new communication technologies such as telephones, mobile devices, email, chat, text messaging, the internet, blogs, websites and interactive videoconferencing to provide psychological services (Siqueira, 2014 cited by Boscariol,2020, page 9).

2.7.1 Advantages and disadvantages of online therapy

At the moment, there are many patients who benefit from this modality. It's also important to note that more and more therapists are using technology as a resource. (www://melhorcomsaude.com.br/).

Advantages

- Level of privacy;
- Offers patient safety;
- Personal comfort;
- Increased range;
- Easy to monitor.

Disadvantages

- Complicated application of psychological tests;
- Communication failure.

2.7.2 Confidentiality and security in communications

There are risks in the improper disclosure of information, messages exchanged in sessions (there is a fear of data being altered in both online and traditional psychotherapy), and it is not possible to say that there are fewer risks in traditional counselling or to guarantee that there are greater risks in online counselling. It is important that traditional chat, email or instant messaging programmes are improved, incorporating encryption technologies in order to increase the security of communications (Midkiff & Wyatt, 2008; cited by Rodrigues, 2016).

2.8 Related applications

2.8.1 TalkSpace

Talkspace is an online and mobile therapy technology solution based in New York City. It was founded by Oren and Roni Frank in 2012. Talkspace users have access to accredited therapists via the website or mobile app on iOS and Android (https://www.talkspace.com). Below is figure 1, the Talkspace app.

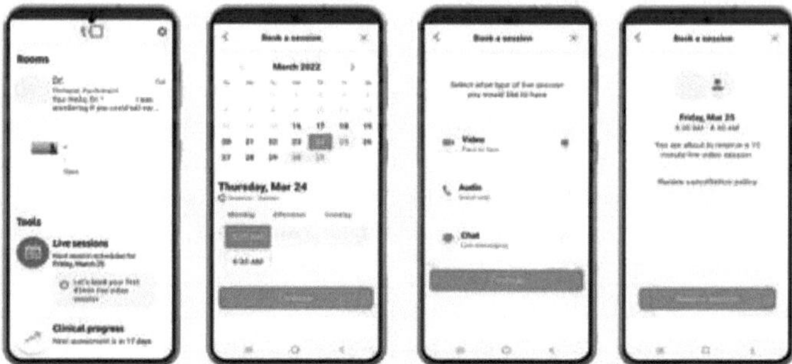

Source: https://www.talkspace.com

2.8.2 Telavite

Telavita is a technological solution for teleconsultations in the health sector that connects health professionals and patients for on-demand remote consultations (https://www.telavita.com.br). Figure 2 below shows the Telavita App.

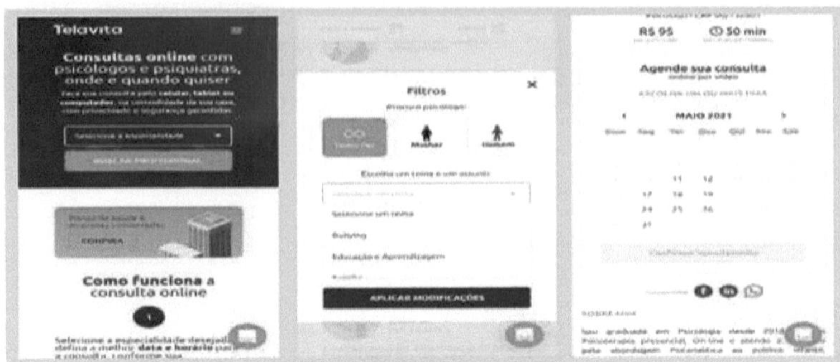

Source: https://www.telavita.com.br

2.8.3 Zenklub

Zenklub is an online psychological counselling platform, where you can book appointments with specialists online without having to leave your home or have a medical insurance. (https://zenklub.com.br/). Figure 3 below shows the Zenklub app.

Source: https://zenklub.com.br

2.8.4 Comparative table of related applications

This is a comparative table designed to illustrate the functionalities of applications with the same characteristics. Table 1 represents a comparative table of the functionalities of mobile applications for online psychotherapy.

Mobile Apps for Online Psychotherapy Consultation				
Features	**SuperMind**	**Talkspace**	**Tel Aviv**	**Zenklub**
Patient Data Management	z	z	z	z
Professional Data Management	z	z	z	z
Diary Management	z	z	z	z
Recover Credentials (Via Email)	z	z	z	z
Exchange Instant Messages	z	z	z	z
Make a Video Call	z	z	z	z
Evaluate Professionals	z	x	x	z
Package Management	z	x	x	z
Statistical Data	z	x	x	x
Link Management (Social Networks)	z	x	x	x
Notification Management	z	x	x	x
Medical Records Management	z	x	x	x

Hardware				
Storage Capacity	20,64 MB	68.07 MB	60.09 MB	56,88 MB

Source: Author, 2023

CHAPTER III

3.DEVELOPMENT METHODOLOGY

The software development process is a set of activities carried out with the aim of obtaining a product through the definition of sub-processes, responsibilities, artefacts and a flow of activity. As organisations become increasingly dependent on the software industry, the problems related to the process of developing these systems become more evident, such as: high cost, high complexity, difficulty in maintenance and frequent disparities between user needs and the product developed.

XP is an agile software development methodology that creates quality systems through a set of values, principles and practices that differ from traditional ones. The XP methodology is recommended for use with small and medium-sized teams to develop software with vague or frequently changing requirements. Its focus is coding and the values that underpin it are simple but efficient communication; simplicity in the design, algorithm and technologies used; feedback on the quality of the code and the progress of the project and the courage to apply changes that arise during development. As it is a test-orientated method, the code is written before the development activity and all the functionalities only have value if they are tested and obtain 100% (one hundred per cent) approval. The XP methodology is suggested because as the code is generated, it must be integrated to avoid incompatibility problems. At the end of the project, the system is delivered when the client is completely satisfied and has nothing more to add in terms of functionality. Other particularities of the XP methodology are programming in pairs, to obtain feedback on the code and encourage the team to make changes when necessary, and the insertion of comments in the code itself, in order to reduce the amount of documents. In addition, communication is free, as long as it is effective. In the XP methodology, there is a greater focus on planning activities to assess the difficulties and feasibility of the project and the validation of integrations takes place in parallel with implementation.

3.1 XP process

Extreme Programming employs an object-oriented methodology, and its development paradigm involves a set of rules and practices within the context of four methodological activities: planning, design, coding and testing (Pressman; Maxim, 2016, p.72).

Planning: this activity, also known as the planning game, begins with the listening process, a requirements elicitation activity that enables the XP team's technicians to understand the intricacies of the software's business environment and obtain a broad view of the requested results, main factors and functionalities. This practice of listening leads to the creation of a set of stories (also known as user stories), which allow the description of the results, characteristics and functionalities requested for the software to be built. The client and the developer work to make joint decisions on certain issues, such as grouping user stories for the next version (software increment) to be developed by the team. Simply put, project velocity is the number of customer stories implemented during the first release. Thus, project velocity can help estimate delivery dates and schedule fulfilment for subsequent releases and determine whether over-commitment has been made to all stories throughout the development project. If overruns occur, the content of the versions is changed and the delivery dates are postponed (Pressman;Maxim,2016,p.73).

Project: the XP project strictly follows the principle. A simple project is always preferable to a more complex representation. In addition, the project provides an implementation guide for a story as it is written; it doesn't get smaller, it doesn't get bigger. The project provides extra functionality because the developer assumes that it will be needed in the future. XP encourages the use of CRC cards as an effective mechanism for thinking about software in an object-orientated context, relevant to the current software increment. If a difficult problem is encountered in the project, as part of a story project, XP recommends creating a working prototype of the project called a point solution, the project prototype is implemented and validated. The aim is to reduce the risks, so that when implementation begins the original estimates for the story containing the project problem are validated.

Refactoring: is the process of changing software in such a way that the external behaviour of the code does not change substantially but the internal structure is improved. Refactoring means improving the coding project after it has been done.

Coding: once the stories have been developed and the preliminary work on the project has been done, the team begins to develop a series of unit tests that will exercise each of the stories to be included in the current version. Once the unit test has been created, the developer can concentrate on the items that will be implemented in order to pass the test. Nothing extraneous is added, the project retains its initial characteristics so as not to lose its essence and quality. Once the code is complete, it can be unit tested, thus making the results available to developers.

Tests: the unit tests created must be implemented using a methodology that enables them. In this way, they will be executed easily and repeatedly. This encourages a regression testing strategy. Whenever the code is changed (which is often, given the nature and philosophy of XP refactoring), XP acceptance tests, also known as customer tests, are detailed by the customer and keep the focus on the features and functionalities of the system, are visible and can be reviewed by the customer. Acceptance tests result from the acquisition of the stories that will subsequently be implemented as part of a software release. Figure 4 below shows the XP process.

Source: Software Engineering - A professional approach

Principles

1. Quick feedback;
2. Simplicity is the best deal;
3. Carry the banner of change / Don't value fear;
4. High quality code;
5. Intense communication.

Values

1. Communication;
2. Simplicity;
3. Feedback;
4. Courage.

3.2 Tools used

Table 2: Tools used.

N°	Tool	Objective
01	Google Form	Statistical Data Collection
02	Trello	Project Task Management
03	MS Project	Designing the Project Timetable
04	Visio	Process Design
05	StarUML	UML Design and Modelling
06	Office 365	Report Preparation
07	Android Studio	Development Environment
08	Java	Programme Language
09	Mysql	Database
10	C#	Back End
11	Postman	Testing WebApi services
12	Figma	Create a prototype

Source: Author, 2023

Figure 5, Tools used.

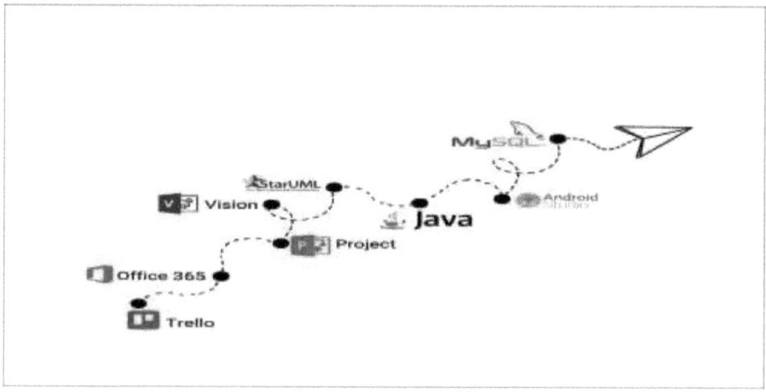

Source: Author, 2023

3.3 Requirements

The requirements of a system describe the actions it must perform, the services it offers, the restrictions and its operation. The requirements reflect the customer's needs for a system whose purpose is determined, such as controlling a device, placing an order or finding information. This process includes a number of elements: analysing, documenting services and constraints. It is considered to be requirements engineering.

3.3.1 Functional requirements

Functional requirements describe the services that the system should provide and how it should react, the input details and under certain circumstances how the system should behave. Functional requirements depend on the nature of the system to be developed and then intended for its possible users and the general approach adopted by the organisation. Below is table 3, functional requirements.

Functional requirements	
RF-01	The application must allow users to register
RF-02	The application should allow patients and professionals to log in
RF-03	The application must be able to recover credentials by sending messages
RF -04	The application should allow you to view the home page
RF -06	The application must be able to visualise
RF -07	The application must allow you to update your profile
RF -08	The app should allow you to search for professionals
RF -09	The application should allow you to create diaries
RF-10	The application must be able to create the patient's file
RF-11	The application must be able to visualise the patient's file
RF-12	The application must be able to make enquiries
RF-13	The application should allow you to create medical reports.

Source: Author,20

3.3.2 Non-functional requirements

These are requirements that are not directly related to the specific services offered by the system to its users. Non-functional requirements can be related to system properties, such as reliability, response time and area occupancy. Non-functional requirements are related to performance, protection and availability, and usually restrict the characteristics of the system as a whole. Table 4 shows the non-functional requirements.

Non-functional requirements	
RNF-01	The application will run in the mobile environment and is compatible with the android operating system.
RNF-02	User-friendly interface, complies with the basic principles of computer ergonomics.
RNF-03	The app is portable and easy to use.
RNF-04	The response time should not exceed 10 seconds.
RNF-05	The application will have a login security mechanism.
RNF-06	Effective access to the application requires checking the e-mail address.

Source: Author, 2023

3.4 Business rule

In order to answer the questions raised, there is an urgent need to create and implement policies to meet the objectives of the business, the needs of the professionals, make the best use of resources and obey the laws or general business conventions. Table 5 below shows the business rules.

N°	Description
RN-01	To access the application, the user must be a Patient or a Professional.
RN-02	To recover your password, you must be a Patient or a Professional without being logged in.
RN -03	To create a profile, you must be a Patient or a Professional without being logged in.
RN -04	Update profile, you need to be a Patient and a Professional and be logged in.
RN -05	To consult professional data, you must be a patient and logged in.
RN -06	To consult patient data, you must be a professional and logged in.
RN -07	To make an assessment, you must be a Patient and logged in.
RN -08	To book an appointment, you must be a Patient and logged in.
RN-09	Create Patient File, you must be a Professional and logged in.
RN-10	To view Patient data, you must be a Professional and logged in.
RN-11	To create notes, you need to be a Professional and logged in.
RN-12	To create a medical report, you must be a Professional and logged in.
RN-13	To view Patient or Professional data, you must be a Patient or Professional and logged in.
RN-14	To make the call, you must be a Professional or Patient and logged in.

Source: Author, 2023

3.5 Use case diagram

To gain a better understanding of the mobile application, the DCU was used to highlight the most relevant elements for a given process. Use cases are the vehicles for capturing requirements and serve as the basis for defining functional requirements. Figure 6 illustrates the functionalities of the application, as well as the human-machine interaction. In this context, the user can only view the data registered by the Patient or Professional. The Patient can register, update their data, view their diary and evaluate the consultation, while the Professional has more resources such as: creating the patient's file, creating the diary and carrying out the consultation.

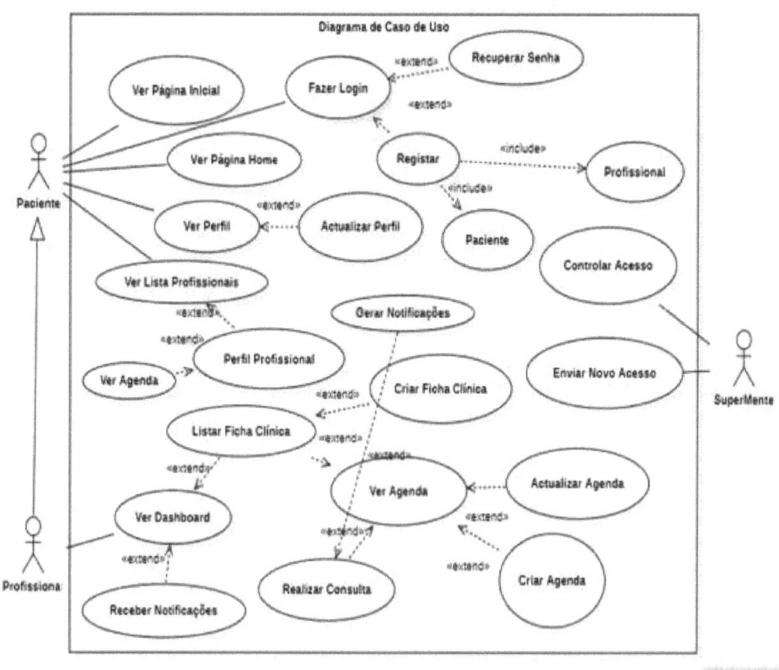

Source: Author, 2023

Table 3: Use case detail

Actors	Use Case	Inclusion	Extension
Patient	■ See home page ■ Register patient ■ Log in ■ Recover password ■ See home page ■ View profile ■ Update profile ■ See agenda ■ Consult a professional ■ Evaluate a professional	Log in Register patient	Register Recover password Update profile Evaluate professional Create notification
Professional	■ See Home ■ Register professional ■ Log in ■ Recover password ■ View Dashboard page ■ View profile ■ Update profile	Register professional	Register Recover password Consult patient Update profile Make an enquiry Receive notification

	■ Consult patient ■ Create a unique file ■ See datasheet ■ Create diary ■ See agenda ■ Make an enquiry		
SuperMind	■ Controlling access ■ Send new access		

Source: Author, 2023

Table 4: Register user

Name	**Register userCSU-01**
Summary	This use case describes the procedures required by the user.
Priority	x EssentialImportantDesirable
Actors: Patient and Professional **Precondition**: The user must fill in the mandatory fields. **Postcondition** : Proceed to the login page.	
Messages	
(MSG-01): The data has been successfully registered. (MSG-02): The data has not been registered. (MSG-03): The fields must be filled in. (MSG-04): These characters are not allowed.	
Main flow	
1. The user requests the registration screen to enter their data. 2. The application displays the user registration screen. 3. The user fills in the data and confirms. 4. The application logs the operation and the use case is closed.	
Alternative flow (1.1)	
1. the user fills in the wrong data and confirms. 2. the system checks and sends a message (MSG-04)	
Flow Exception (1.1):Failure to fill in mandatory data	
1. the user does not fill in any data. 2. the application issues a message (MSG-03).	
Business rules	
RN-03.	

Source: Author, 2023

Name	**Log inCSU-02**
Summary	This use case describes the procedures required to log in.
Priority	x EssentialImportantDesirable
Actors: Patient and Professional **Precondition**: Fill in your email address and valid password. **Post-Conditioning** : Home page access	
Messages	
(MSG-01): Welcome. (MSG-02): This data does not match, please try again. (MSG-03): The fields must be filled in.	

Main flow
1.The user requests the login screen. 2. the application displays the login screen. 3) The user enters their email address and password. 4.The application checks that the data entered is valid. 5. the application displays the home page.
Alternative flow (1.1)
1. The user fills in the unauthorised data and asks for confirmation. 2.The application checks the data and sends a message (MSG-02) and returns to point 3 in the main flow.
Flow Exception (1.1):Invalid User or Password
a) The user does not fill in any data. b) The application issues a message (MSG-03).
Business rules
RN-01.

Source: Author, 2023

Table 6: Recover password

Name	**Recover passwordCSU-03**
Summary	This use case describes the procedures for recovering the user's password.
Priority	Essential x ImportantDesirable
Actors: Patient and Professional **Precondition**: The user must fill in the mandatory fields. **Postcondition** : Proceed to the login page.	
Messages	
(MSG-01): The data was sent successfully. (MSG-02): Data was not sent, try again. (MSG-03): The fields must be filled in. (MSG-04): These characters are not allowed.	
Main flow	
1. The user requests a recovery screen to add their data. 2. The application displays the password recovery screen. 3. The user fills in the data and confirms. 4. The application logs the operation and the use case is closed.	
Alternative flow (1.1)	
1. the user fills in the wrong data and confirms. 2.the application checks and sends a message (MSG-04)	
Flow Exception (1.1):Failure to fill in mandatory data	
1. the user does not fill in any data. 2. the application issues a message (MSG-03).	
Business rules	
RN-02.	

Source: Author, 2023

Name	**Send credentialsCSU-04**

Summary	This use case describes the procedures for recovering credentials.
Priority	x EssentialImportantDesirable
Actors: Patient and Professional **Precondition**: The user cannot be logged in. **Post-Condition** : Display whether or not the credentials have been sent.	
Messages	
(MSG-01): Message sent successfully. (MSG-02): Message was not sent, please try again.	
Main flow	
I. Nothing.	
Alternative flow (1.1)	
I. Nothing.	
Flow Exception (1.1):Failure to fill in mandatory data	
I. Nothing.	
Business rules	
I. Nothing.	

Source: Author, 2023

Name	See home pageCSU-05
Summary	This use case describes the procedures for visualising the home page.
Priority	EssentialImportantxDesirable
Actors: Patient and Professional **Precondition**: The user must be logged in. **Post-Condition** : View the Home page.	
Messages	
(MSG-01): Please wait while system processes.	
Main flow	
1. The user requests the main page screen. 2. The application displays the main page screen.	
Alternative flow (1.1)	
I. Nothing.	
Flow Exception (1.1):Failure to fill in mandatory data	
I. Nothing.	
Business rules	
RN-01.	

Source: Author, 2023

Name	View profileCSU-06
Summary	This use case describes the procedures for visualising the profile.
Priority	EssentialImportantxDesirable
Actors: Patient and Professional **Precondition**: The user must be logged in. **Post-Condition**: View the profile.	
Messages	
(MSG-01): Please wait while system processes.	
Main flow	
1. The user requests the profile page screen. 2. The application displays the profile page screen.	

Alternative flow (1.1)
I. Nothing.
Flow Exception (1.1):Failure to fill in mandatory data
I. Nothing.
Business rules
RN-13.

Source: Author, 2023

Name	**Update profileCSU-07**
Summary	This use case describes the procedures required to update the user's profile.
Priority	x EssentialImportantDesirable
Actors: Patient and Professional **Precondition**: The user must be logged in. **Post-Condition** : The user must fill in all the fields.	
Messages	
(MSG-01): The data has been successfully updated. (MSG-02): The fields must be filled in. (MSG-03): These characters are not allowed.	
Main flow	
1. The user requests the inclusion of their data. 2. The application displays the user registration screen. 3. The user fills in the data and confirms. 4. The application logs the operation and the use case is closed.	
Alternative flow (1.1)	
1. the user fills in the wrong data and confirms. 2.the application checks and sends a message (MSG-03)	
Flow Exception (1.1):Failure to fill in mandatory data	
1. the user does not fill in any data. 2. the application issues a message (MSG-02).	
Business rules	
RN-04.	

Source: Author, 2023

Name	**Create agendaCSU-08**
Summary	This use case describes the procedures required to create the appointment diary.
Priority	x EssentialImportantDesirable
Actor: Patient **Precondition**: The user must be logged in. **Post-Condition** : Visualise the diary created.	
Messages	
(MSG-01): The appointment was created successfully. (MSG-02): The fields must be filled in. (MSG-03): These characters are not allowed.	
Main flow	
1. The patient requests the Create diary screen to add data.	

2. The application displays the create diary screen. 3. The patient fills in the details and confirms. 4. The application logs the operation and the use case is closed.
Alternative flow (1.1)
1. the patient fills in the wrong data and confirms. 2.the application checks and sends a message (MSG-03)
Flow Exception (1.1):Failure to fill in mandatory data
1. the patient does not fill in any data. 2. the application issues a message (MSG-02).
Business rules
RN-08.

Source: Author, 2023

Name	**Consult a professionalCSU-09**
Summary	This use case describes the procedures required to consult a professional.
Priority	EssentialImportantxDesirable
Actor: Patient **Precondition**: The user must be logged in and fill in the enquiry field. **Post-Condition** : View the list of professionals.	
Messages	
(MSG-01): Please wait while the system processes. (MSG-02): The data does not match. (MSG-03): The field must be filled in.	
Main flow	
1. The patient requests the consult a professional screen. 2. The application displays the consult a professional screen. 3. The patient fills in the details and confirms. 4. The application logs the operation and the use case is closed.	
Alternative flow (1.1)	
1. the patient fills in the wrong data and confirms. 2.the application checks and sends a message (MSG-03)	
Flow Exception (1.1):Failure to fill in mandatory data	
1. the patient does not fill in any data. 2. the application issues a message (MSG-02).	
Business rules	
RN-05 and 07.	

Source: Author, 2023

Name	**Create patient recordCSU-10**
Summary	This use case describes the procedures required to create a patient record.
Priority	x EssentialImportantDesirable
Actor: Professional **Precondition**: The user must be logged in. **Post-Condition** : Visualise the diary created.	

Messages
(MSG-01): The data has been successfully registered. (MSG-02): The fields must be filled in. (MSG-03): These characters are not allowed.
Main flow
1. The professional requests the create patient record screen to add data. 2. The application displays the create diary screen. 3. The professional fills in the details and confirms. 4. The application logs the operation and the use case is closed.
Alternative flow (1.1)
1) The professional fills in the wrong data and confirms. 2. the application checks and sends a message (MSG-03)
Flow Exception (1.1):Failure to fill in mandatory data
1. the professional does not fill in data. 2. the application issues a message (MSG-02).
Business rules
RN-09.

Source: Author, 2023

Name	**See patient fileCSU-11**
Summary	This use case describes the procedures for viewing the patient's file.
Priority	EssentialImportantxDesirable
Actor: Professional **Precondition**: The user must be logged in. **Post-Condition**: View the patient's file.	
Messages	
(MSG-01): Please wait while system processes.	
Main flow	
1. The professional requests the screen page of the patient's file. 2. The application displays the patient record page.	
Alternative flow (1.1)	
I. Nothing.	
Flow Exception (1.1):Failure to fill in mandatory data	
I. Nothing.	
Business rules	
RN-06 and 10.	

Source: Author, 2023

Name	**Create notesCSU-012**
Summary	This use case describes the procedures required by the user.
Priority	x EssentialImportantDesirable
Actor: Professional **Precondition**: The user must fill in the required fields. **Post-Condition**: Visualise the data entered.	
Messages	
(MSG-01): The data has been successfully registered. (MSG-02): The data has not been registered.	

(MSG-03): The fields must be filled in. (MSG-04): These characters are not allowed.
Main flow
1. The professional requests the notes screen to add their data. 2. The application displays the notes screen for the consultation. 3. The professional fills in the details and confirms. 4. The application logs the operation and the use case is closed.
Alternative flow (1.1)
1. the professional fills in the wrong data and confirms it. 2.the application checks and sends a message (MSG-04)
Flow Exception (1.1):Failure to fill in mandatory data
1. the professional does not fill in data. 2. the application issues a message (MSG-03).
Business rules
RN-03.

Source: Author, 2023

Table 16: Making video calls, receiving and sending messages

Name	**Making video calls, receiving and sendingCSU-13 messages**
Summary	This use case describes the procedures for making a video call, receiving and sending messages between psychologist and patient.
Priority	EssentialImportantxDesirable
Actors: Professional and Patient **Precondition**: The user must be logged in. **Post-condition**: Audiovisual message sharing.	
Messages	
(MSG-01): Please wait while the application processes.	
Main flow	
1. The user requests the message screen. 2. The application displays the message screen. 3. The sender types and sends the message. 4. The receiver confirms the message.	
Alternative flow (1.1)	
1. The user requests the video screen called up. 2. The application displays the video call screen. 3. The sender triggers the call. 4. The receiver confirms the call.	
Flow Exception (1.1):Failure to fill in mandatory data	
I. Nothing.	
Business rules	
RN-14.	

Source: Author, 2023

Table 17: Login test case

Name	**Login test case**	**TC-01**
Summary	This test case validates the steps required to access the application.	

Messages			
N°	CT-Email	CT-Password	Results
1	Nil	OK	Error
2	OK	Nil	Error
3	OK	OK	Success
Note: All fields must be filled in.			
Use case to validate			
CSU-02			

Source: Author, 2022

Table 18: Test case-register user

Name	Test case -Register user				TC-02
Summary	This test case validates the steps required to register a user in the system.				
Messages					
N°	CT-Name	CT-Appealed	CT-Email	CT-Password	Results
1	Nil	OK	Nil	OK	Error
2	OK	Nil	OK	Nil	Error
3	OK	OK	OK	OK	Success
Note: All fields are mandatory				torio.	
Use case to validate					
CSU-01					

Source: Author, 2023

Table 19: Test case - recover password

Name	Test Case -Recover password		TC-03
Summary	This test case validates the steps required to access the application.		
Messages			
N°	CT-Email	CT-Password	Results
1	Nil	OK	Error
2	OK	Nil	Error
3	OK	OK	Success
Note: All fields must be filled in.			
Use case to validate			
CSU-03			

Source: Author, 2023

3.6 Activity diagram

The activity diagram represents a set of actions that can be carried out according to the user's needs and the application's availability. Figure 7 illustrates the procedures that can be carried out by the patient or professional while handling the application.

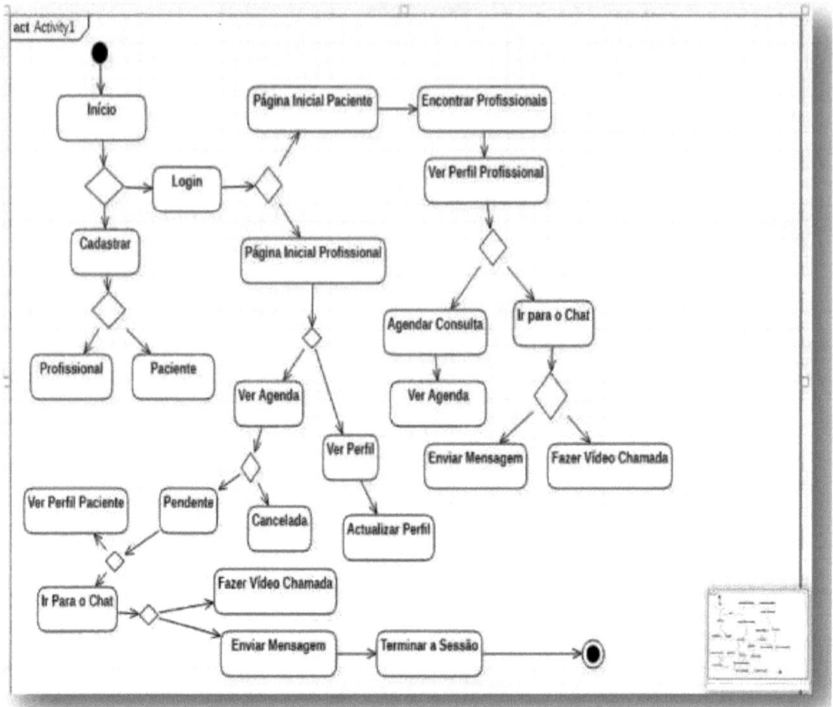

Source: Author, 2023

3.7 Class diagram

Figure 8 illustrates the logical relationship between entities that have been implemented in the application. In the relationship between entities, some terms have been used, such as: has, concerns, validates and does. These terms help us to understand the relationship between one entity and another. The expression concerns means that there is a relationship between entities of the type: one-to-one; the expressions has, does, validates and fills means that there is a relationship between entities of the type: one-to-many.

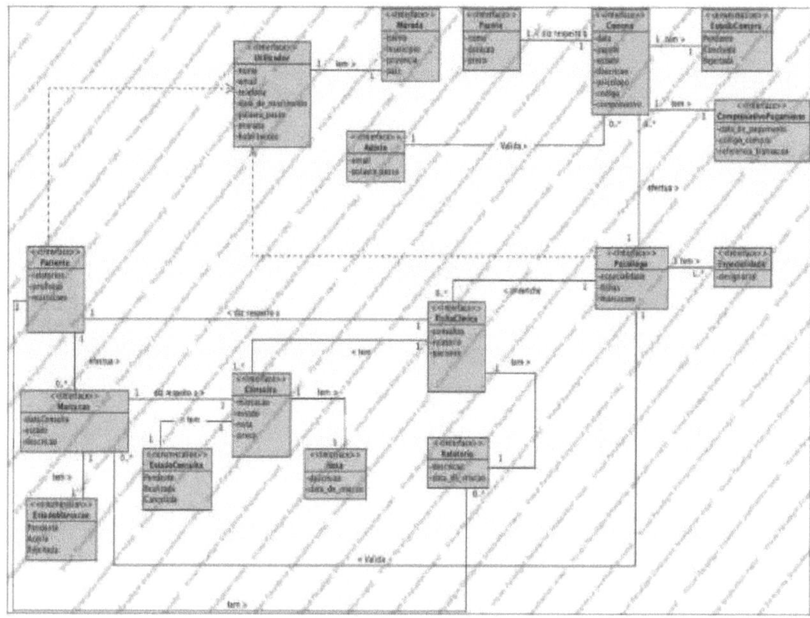

Source: Author, 2023

3.8 Database diagram

Figure 9 illustrates how logical data modelling is carried out to design the DB schema at a technology-independent level. The logical data model is seen as a refinement of the conceptual model by introducing details, namely: the description of attributes, the description of the data type or the explanation of the multiplicity constraints of its associates. Logical models are used to enable database developers to transform them into relational schemas.

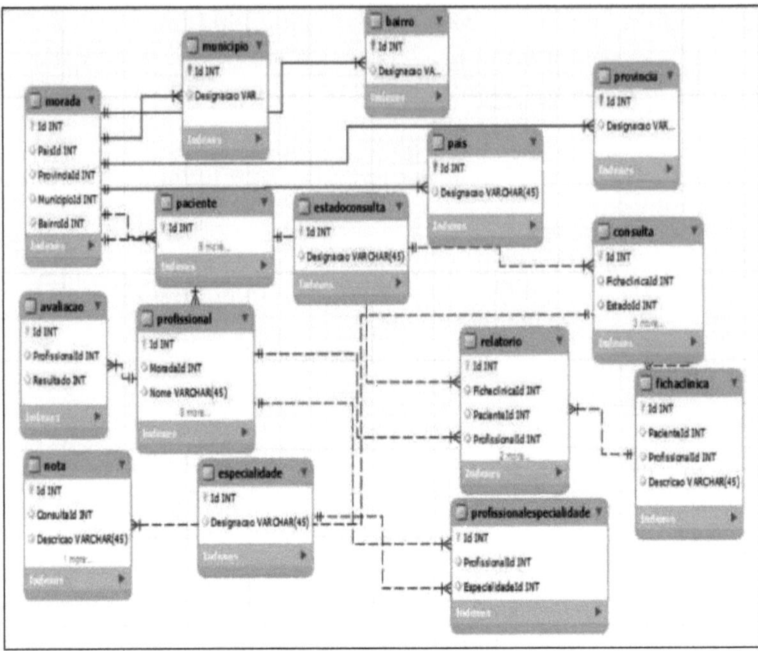

Source: Author, 2023

3.9 Sequence diagram

Figure 10 illustrates the validation of the procedures carried out when updating profile data. The user makes a request, the application responds, then the user enters the data, confirms and returns to the starting point.

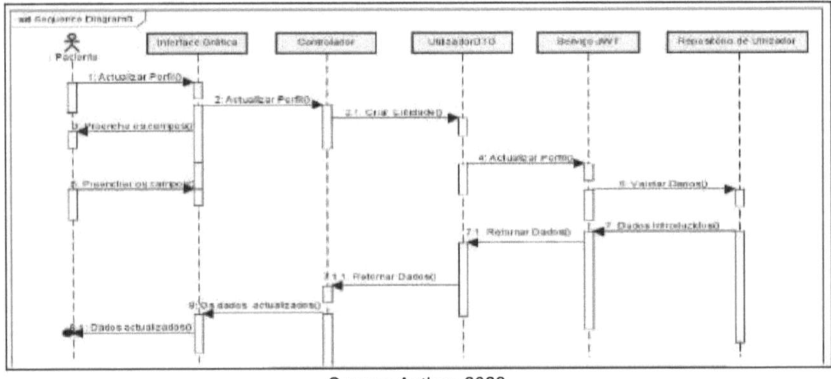

Source: Author, 2023

Figure 11 illustrates the communication between the mobile application and the Web application. The Web application provides the *endpoints*, which will be consumed by the client (mobile application) through the WebApi that will be responsible for brokering customer orders.

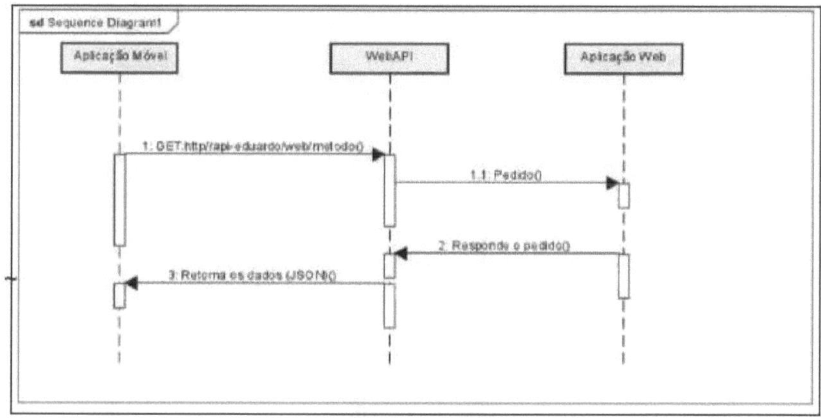

Source: Author, 2023

3.10 System architecture

System architecture refers to the overall organisation of software and the way in which architecture provides conceptual integrity for a system. Architecture is the organisation of components that can be generalised to represent the main elements of a system and their interactions. The webApi is the backend where the entire structure of the project is contained, this application communicates directly with the database whenever it receives any request. The Mobile App is the mobile application that consumes the data provided by the WebApi via an endpoint. App Logs are records of events that occur in the application. This architecture represents the organisation from which the more detailed activities of the project are conducted. Figure 12 illustrates the system's architecture.

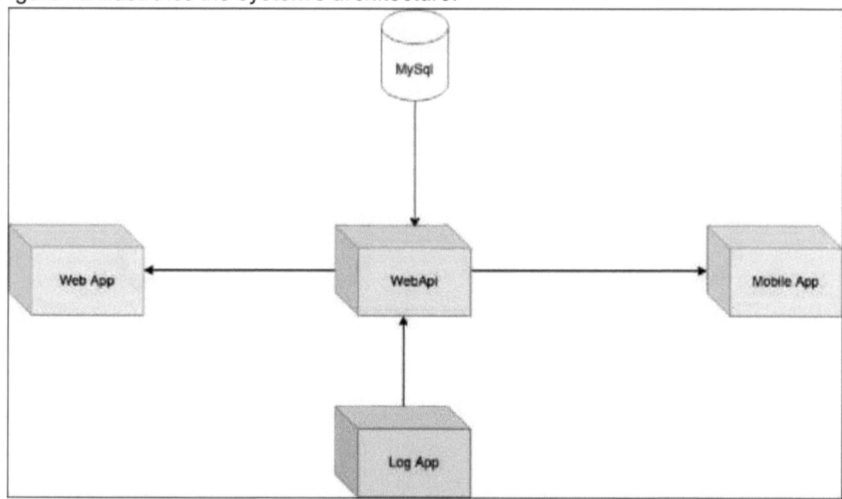

Source: Author, 2023

3.11 Planning

Figure 13 illustrates the planning phase, with the login page on the left and the create account page on the right.

Source: Author, 2023

Figure 14 illustrates the planning phase, the left-hand side of the home page and the right-hand side of the find professionals page.

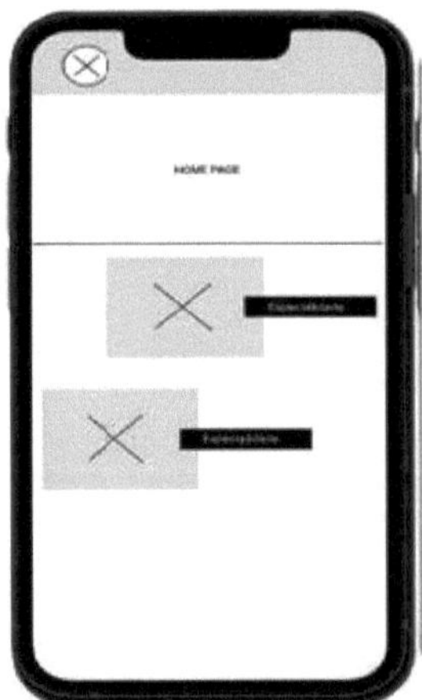

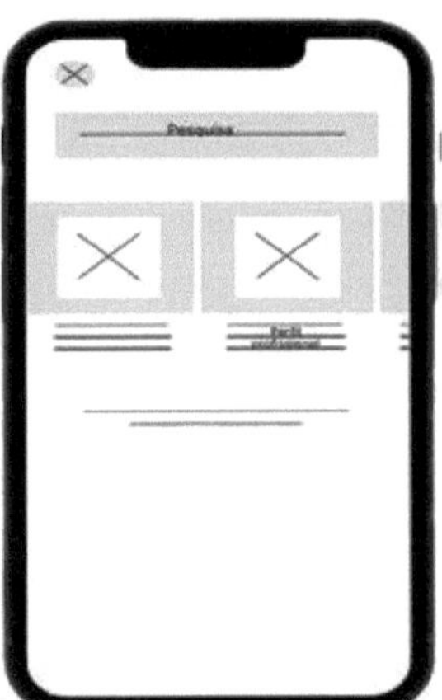

Source: Author, 2022

Figure 15 illustrates the planning phase, with the profile page on the left and the professional's desktop page on the right.

Source: Author, 2023

Figure 16 illustrates the planning phase of the page that makes it possible to realise the video conference call.

Source: Author, 2023

3.12 Project

Figure 17 illustrates the design phase of the login page. In this phase, the project shows a real image compared to the planning phase.

Source: Author, 2023

Figure 18 illustrates the project phase, with the professional registration page on the left and the patient registration page on the right. In this phase, the project shows a real picture compared to the planning phase.

Figure 19 illustrates the project phase of the home page. In this phase, the project shows a real picture compared to the planning phase

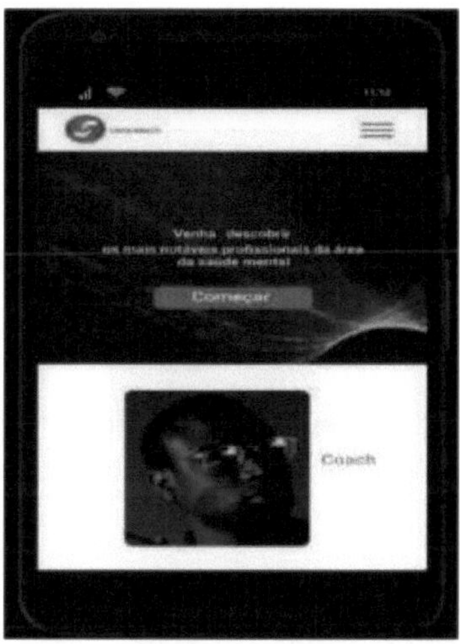

Source: Author, 2023

Figure 20 illustrates the design phase of the Find Professional page. In this phase, the project shows a real picture compared to the planning phase.

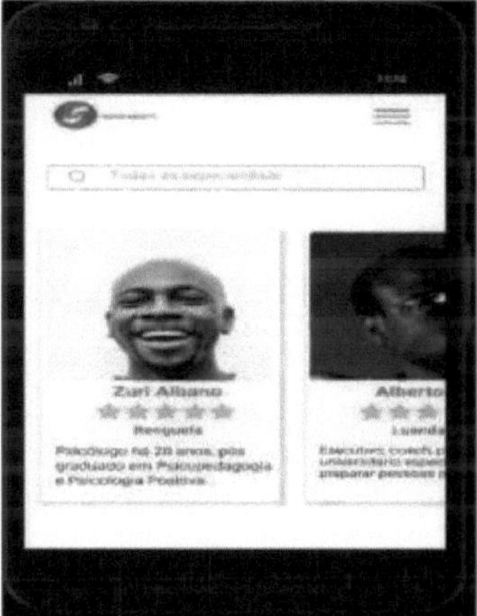

Figure 21 illustrates the design phase of the create diary page. In this phase, the project shows the real picture more closely than in the planning phase.

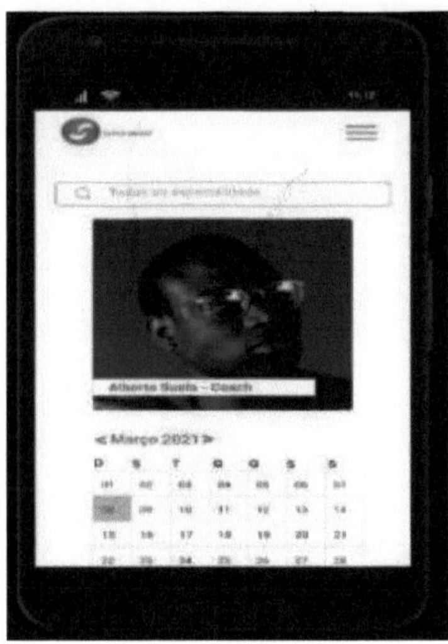

Source: Author, 2023

Figure 22 illustrates the design phase of the video conference call page. In this phase, the project shows a real image compared to the planning phase.

CHAPTER IV

4.PRESENTATION AND ANALYSIS OF RESULTS

Figure 23 illustrates the main page of the application, which contains important information about the procedures that can be carried out for both the user and the administrator. In addition to the information, the user can see various specialities that can be found in the application.

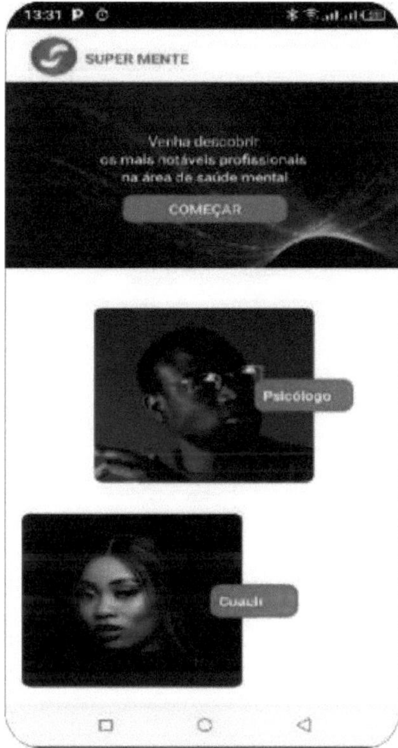

Figure 24 shows the login screen, with the create account *link*, forgot your password *link* and the log in button. Before logging in, the user must fill in their credentials and then click on the log in button. When this is done, the data is sent to the *backend*, which will check whether the data entered matches or not via the database. If the credentials are true, the patient is sent to the application's *homepage*; if they are a professional, they are sent to the *dashboard*.

Figure 25 shows the registration screen. This area is reserved for professionals. Before accessing the system, professionals must first register in order for their credentials to be saved. This process is carried out by filling in all the fields and is validated by clicking on the create account button, as shown in the figure below.

Source: Author, 2023

Figure 26 shows the registration screen and an area reserved for patients. Before accessing the system, patients are required to register for the first time in order for their credentials to be saved. This process is carried out by filling in all the fields and is validated by clicking on the create account button.

Source: Author, 2023

Figure 27 illustrates the *homepage,* a space reserved for the patient to search for professionals. The patient can see a variety of specialities, as well as the procedures to take into account if they wish to carry out certain tasks in the application. The user can click on the top right-hand corner to access the menu, which contains various *links* that can be accessed.

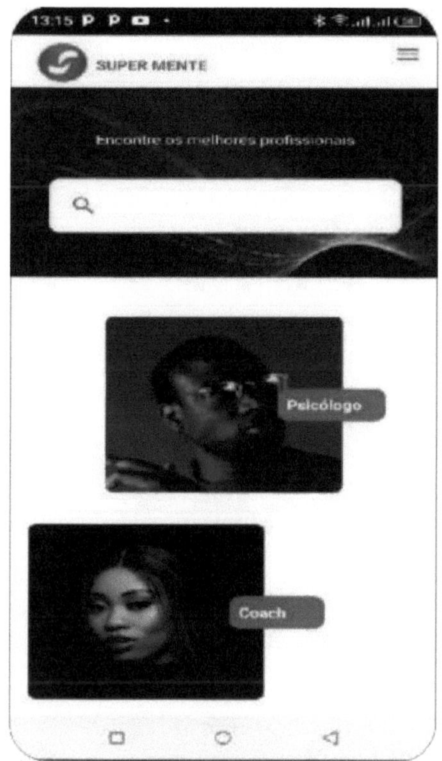

Source: Author, 2023

Figure 28 illustrates the menu where the professional can navigate by clicking. The menu allows the user to carry out the action they want. By clicking on one of the *links*, the application produces an event and the user is directed to one of the screens represented in the menu.

Source: Author, 2023

Figure 29 illustrates the waiting time after the user has requested the list of professionals and can choose for a possible consultation. The figure below shows how the application is loading the data, once the user has made the request. In addition to the space that will be filled with the list of professionals registered in the application, below is information that can direct the user.

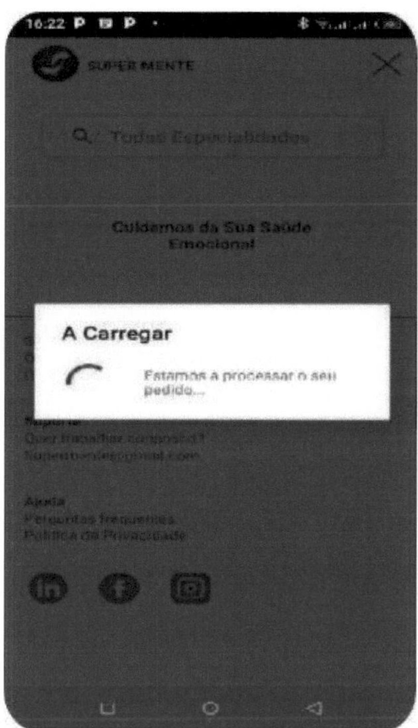

Figure 30 illustrates a space where there is a huge list of professionals who have an assessment, carried out on the basis of the service provided. The patient has the opportunity to choose the professional. The application has this functionality to allow the patient to have prior knowledge of a particular person. There is a history that can make it easier for the professional to be contacted or not. This element can also contribute to the quality of the service provided, as well as the credibility of the application itself.

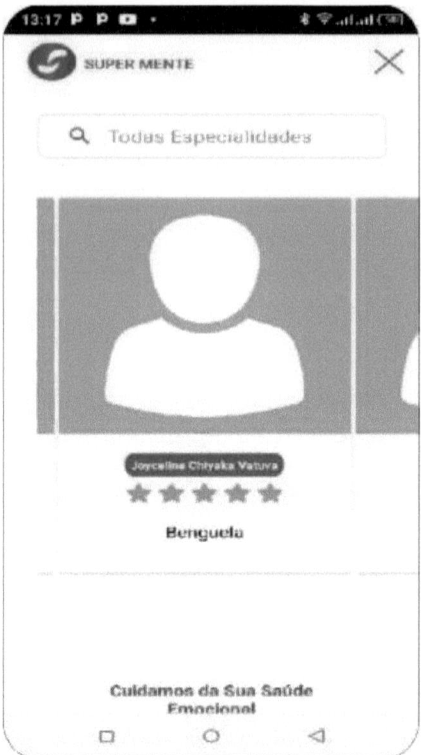

Figure 31 illustrates the space where the professional is evaluated according to the patient's experience of the previous consultation. The evaluation is made by *clicking* on a star and then the evaluation must be sent to be saved in the database.

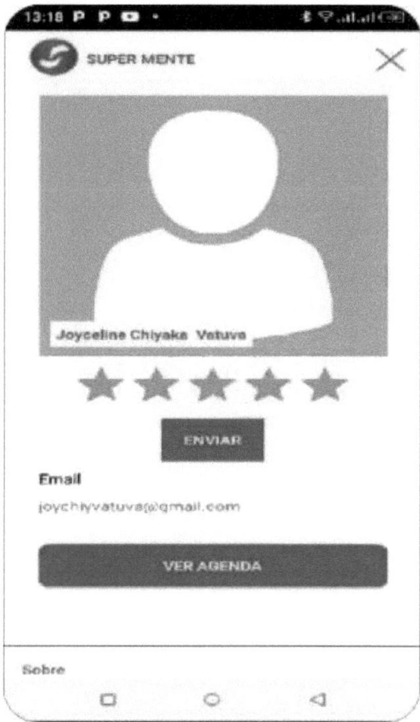

Figure 32 shows the *dashboard,* a space reserved for the professional to navigate the application via the menu. In addition to the menu on the top right-hand side, there are also buttons that can direct the professional to the screen where they want to carry out a particular task. On the professional's desktop it is possible to view statistical data expressed as a percentage that reflects a set of consultations carried out and cancelled.

Figure 33 shows the list of patients who have already been seen by a specialist. The professional can access the patient's profile by *clicking on* the view button, *clicking* on the name and *clicking* on the image. On this screen, it is possible to visualise some of the patient's data which can help the professional if they wish to carry out another consultation. The professional can scroll through the list of patients and select who has booked an appointment. The button below allows you to create a clinical record by clicking on it and filling in the modal and when you click on save, the professional's details are displayed on the screen.

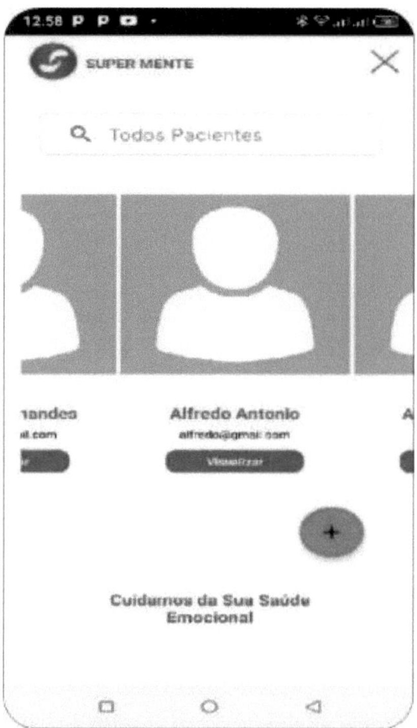

Figure 34 illustrates the professional's profile. In this profile, in addition to the various fields that exist and relate to the professional's data, we have highlighted the name, surname, e-mail address and the areas in which the professional works. In the profile, professionals have the opportunity to make changes to their details as required. To do this, they fill in the fields they want to change and then *click on* save. In turn, the application will validate and forward it to the database, whereupon the new data will be updated. In the case of the area of specialisation, if the professional wants to reduce it, all they have to do is *click on* the name of the speciality and it will check or uncheck it and then save the changes.

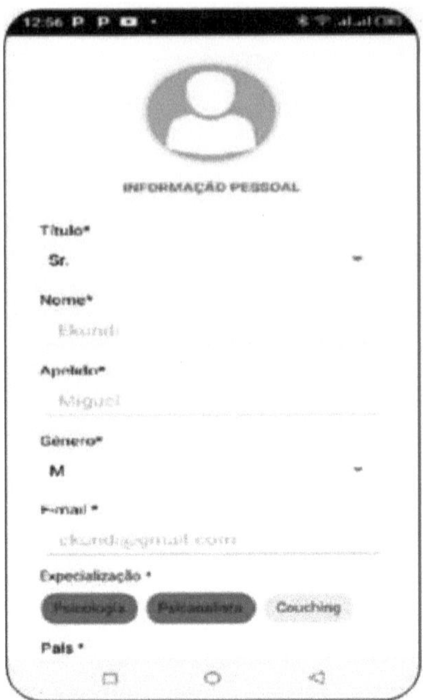

Source: Author, 2023

Figure 35 illustrates the appointment space. This space is reserved for patients and professionals to check their diaries. At the top, it shows the patient's details, followed by a list of pending appointments, appointments made and appointments cancelled. Then the calendar with the updated date. This administrative procedure works as follows: the patient notifies the professional to make an appointment, the professional informs the patient of their availability, if the patient agrees, the professional makes the appointment and notifies the patient that the appointment has been made, on the date and time, according to the conversation. In turn, the patient can view this information from their profile.

Figure 36 illustrates the environment where you can multitask, such as logging out, updating the diary, cancelling the diary and making an enquiry via a video conference call. To view the multitasking modal, simply click on one of the items. This operation is carried out exclusively by the professional, as they define when and at what time they are available to provide the service.

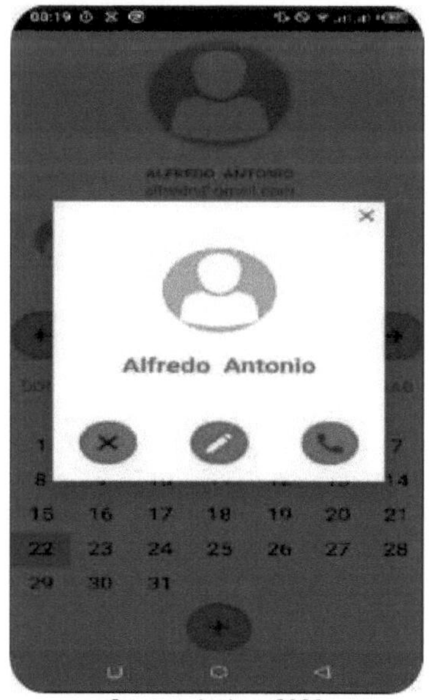

Source: Author, 2023

Figure 37 illustrates the space where the consultation can be carried out by video conference. This reserved space allows interaction between the professional and the patient. The sharing of information in person via audio visual devices, in which the parties are involved in an interactive environment, the patient relates certain experiences and the professional makes the necessary notes.

Figure 38 illustrates the space for interaction between the professional and the patient via instant messaging. The patient or professional is free to use the *chat* function if there is any interference with the live communication. This feature makes it possible to maintain the continuity of the consultation so that the data obtained can be evaluated in order to make decisions.

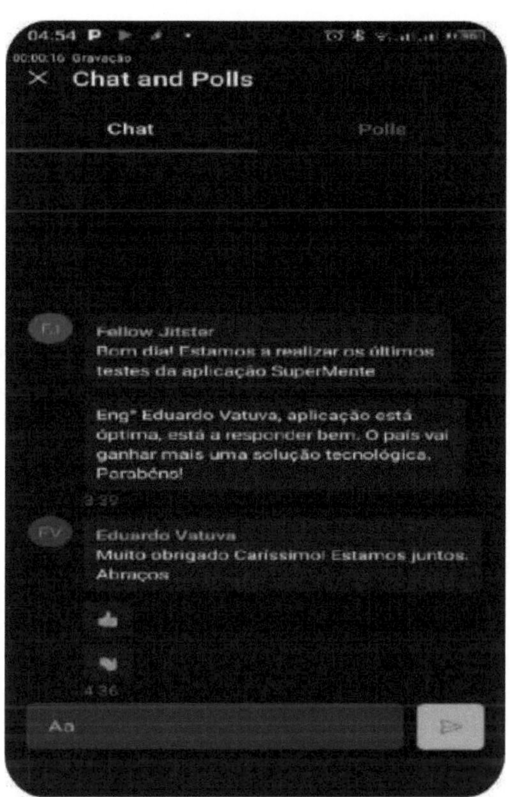
04:54
00:00:16 Gravação
Chat and Polls
Chat
Polls
Fellow Jitster
Bom dia! Estamos a realizar os últimos testes da aplicação SuperMente
Engº Eduardo Vatuva, aplicação está óptima, está a responder bem. O país vai ganhar mais uma solução tecnológica. Parabéns!
3:39
Eduardo Vatuva
Muito obrigado Caríssimo! Estamos juntos. Abraços
4:36
Aa

CHAPTER V

5.CONCLUSIONS

Once the subject has been addressed, the development of a mobile application for intervention in cognitive behavioural psychotherapy. There is an urgent need to state that this study is not finished, given its complexity, and it is necessary to continue in the search for knowledge in order to better understand it. The conclusion is that

In the fundamental theory, it was found that given the nature of the application, it is necessary to adjust some functionalities based on the Angolan legal system. On the other hand, the scope of the online cognitive behavioural psychotherapy service is not limited to Angola.

Planning has made it possible to create certain processes aimed at reducing the bureaucratic administrative procedures that still exist in the traditional service model.

The project defined the application's requirements, which served as a guiding thread for the project's conception. At this stage it was possible to visualise the real image of the application, which helped to bring theory and practice into line.

The test allowed us to verify the functionality of the application through science and technology. The SuperMente application is fully functional, without ruling out the possibility of errors.

5.1 Recommendations

The work is not a finished product, so here are some important recommendations for the materialisation of the project. The application is functional and answers the problems presented, such as:

- The application can be used in private and public institutes;
- Psychological professionals have an alternative way of providing services remotely;
- Patients are free to choose the specialist they want.

BIBLIOGRAPHICAL REFERENCES

How do I get Java for Mobile Devices?: https://www.java.com/pt-BR/download/help/java_mobile_en-br.html/2022, accessed on 24 February 2022.

CORREA, F.(2013) Gestor de Psicologia: Gestão de Consultios de Psicologia. Monograph for the Regional University of Blumenau.

GIL R. L.(2008) Licenciatura em Ciencias Biologicas, Disciplina de Pesquisa do Ensino de Ciencias e Biologia, Tipos de Pesquisa.

GONQALVES, S. T.; Belmino, M. C.B.O.(2017) Virtual Environment as Professional Space: online psychological services. Revista Interfaces do Centro Universitario Dr. Leao Sampaio - UNILEAO^, v. 4, n. 2.

HERMOSILLA, L. (2008) Psychotherapy and the Internet: distance therapy. Revista cientffica electronica de psicologia - issn: 1806-0625. Year V - Number 08 - February. https://docs.google.com/forms/d/1ceQfGdMh8-ZtEocnvbCRNRz7wh_fn-IrtrW6pgwqijY/edit accessed, 05 October 2021.

https://br.financas.yahoo.com/noticias/como-funciona-o-zenklub-plataforma accessed 24 February 2022.

https://images.app.goo.gl/BHnUWWy66xgYAoPT7 accessed 25 February 2022.

https://images.app.goo.gl/HL6PUKVMqd8YBYDX6 accessed 25 February 2022.

https://images.app.goo.gl/X1i1jkhpzCjHZshn9 accessed 25 February 2022.

https://www.talkspace.com/ accessed on 24 February 2022.

https://zenklub.com.br/busca/?utm_source=google&gclid=EAIaIQobChMItLrYjN-0-AIVBwGLCh3gNQ6pEAAYASAAEgLP5fD_BwE accessed 25 February 2022.

LEITE, N. J. S. (2016).Psicoterapia Online: uma possibilidade de actua^ao para psicologia na era digital. Monograph from the Faculty of Education and Environment - Faema, Ariquemes-ro.

Healthy mind, Advantages and disadvantages of online therapy: https://melhorcomsaude.com.br/terapia-online-vantagens-desvantagens/,2021 accessed 04 January 2022.

MUNHOZ,J. L.(2019). Online Psychological Care: A new concept in psychotherapy. XVII Campos Gerais Scientific Conference. Ponta Grossa, 23 to 25 October.

What is Java technology and why do I need it?:https://www.java.com/pt-BR/download/help/whatis_java.html/,2022, accessed 24 February 2022. PINHATI, M. M.(2013). Internet therapy: limits and possibilities in the perception of psychotherapists. Monograph from the Federal University of Rio Grande do Sul, Porto Alegre, 2013.

Pressman R;Maxim B.(2016). Software Engineering: A Professional Approach, 8th edition, AMGH, Editora Ltda.

Telavita and a telecare company: https://www.telavita.com.br, accessed 24 February 2022.

Translated from English into Portuguese: Talkspace is an online therapy company: https://en.wikipedia.org/wiki/Talkspace, accessed 24 February 2022.

RODRIGUES, C. G.; TAVARES, M.A.(2016). Online psychotherapy: growing demand and suggestions for regulation. Scientific Article. Psicologia em Estudo, vol. 21, num. 4, octubre-diciembre, 2016, Universidade Estadual de Maringa, Brazil.

ROMANO, J. H.; SANTOS, C.N; ALVES V. A. F.(2019). Online Psychotherapy and its Challenges in Post-Modernity. Monograph from the Salesiano Auxilium Catholic

University Centre, Psychology Course.
ULKOVSKI, E. P.; DA SILVA, L. P. D.; RIBEIRO, A. B.(2017). Online psychological care: current perspectives and challenges for psychotherapy. Revista de IniciaQao Cientica da Universidade Vale do Rio Verde, v. 7, n.

APPENDIX

APENDICES

GREGORIO SEMEDO UNIVERSITY MASTER'S IN ENGINEERING INFORMATICS

Appendix I - Questionnaire for psychology professionals

1. What benefits does online consultation offer?
2. What tools do you consider essential for carrying out your work?
3. On average, how long would a face-to-face consultation take?
4. What is your speciality?

Appendix II-What is the benefit of online consultation?

Figure 1

Que benefícios tem a consulta online ?

51 respostas

Ser feita apartir de qualquer lugar ou distância desde que se tenha rede

Salva vida

É pratico, não há a necessidade de deslocação e pode ser feita a partir de casa ou serviço.

Proteger nos do Corona virus

Reencontrar-se com o eu.

Facilita devido a situação que vivemos e não teremos que nos deslocar.

Tem vários benefícios, ela pode ser realizada a partir de qualquer momento e distante do avaliador. Ajuda-nos a estar mais preparado e mais calmo

What benefits does the online consultation offer?
51 answers
It can be done from anywhere, as long as you have a network.
Life-saving
It's practical, there's no need to travel and it can be done from home or work.
Protect us from the Corona virus
Finding yourself again.
It's easier because of the situation we're in and we won't have to move.
It has many benefits: it can be done at any time and away from the assessor. It helps us to be more orientated and calmer.

Source: Author, 2022

Appendix III-What tools do you consider essential for carrying out your work?

Figure 2

Que ferramentas considera essencial para o desempenho da sua atividade laboral ?

47 respostas

Se for online, um aparelho ligado a net com app para video chamada e privacidade.

Formulários e testes.

Licenciatura, formação ética e Especialização

Questionário, entrevista.

TCC, TESTES PSICOLÓGICOS ,DOSE ,PASSOS DOS ALCOÓLICOS ANONIMOS, E NARCÓTICOS ANÓNIMOS E TERAPIA EMDR.

O Computador é uma ferramenta essencial para desepenhar a minha actividade laboral.

Computador e telefone

Interação interpeaoal

Maior inserção e abrangência das Tics (tecnologias de informação

What tools do you consider essential for carrying out your work?
47 replies
If it's online, a device connected to the net with an app for video calls and privacy.
Formularies and tests.
Degree, ethical training and specialisation
Questionnaire, interview.
TCC, PSYCHOLOGICAL TESTS .DOSE .STEPS OF ALCOHOLICS ANONYMOUS. AND NARCOTICS ANONYMOUS AND DRUG THERAPY.
O A computer is an essential tool for carrying out my work.
Computer and telephone
Interpeonal interaction
Greater inclusion and coverage of ICT (information technology)

Source: Author, 2022

Figure 3

Em média quanto tempo levaria uma consulta ?

53 respostas

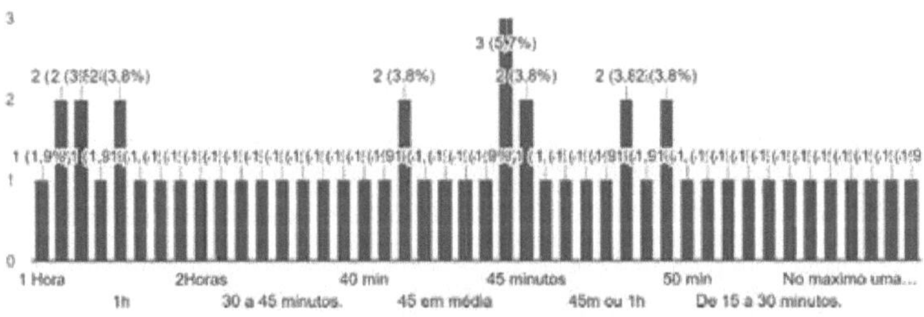

Source: Author, 2023

Appendix V-What is your speciality?

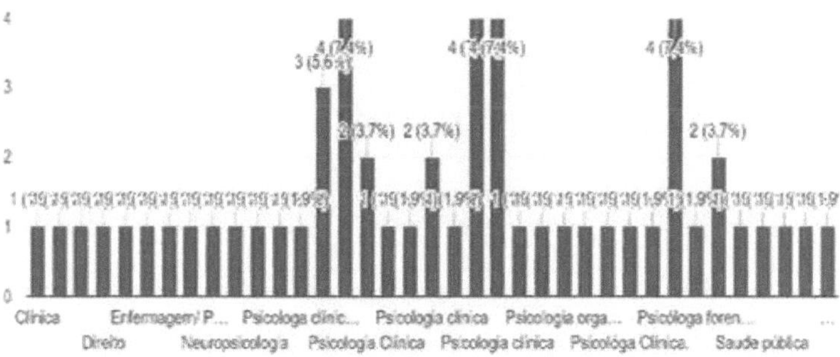

Source: Author, 2023

Figure 4

ANNEXES

Annex I - Population interviewed by gender

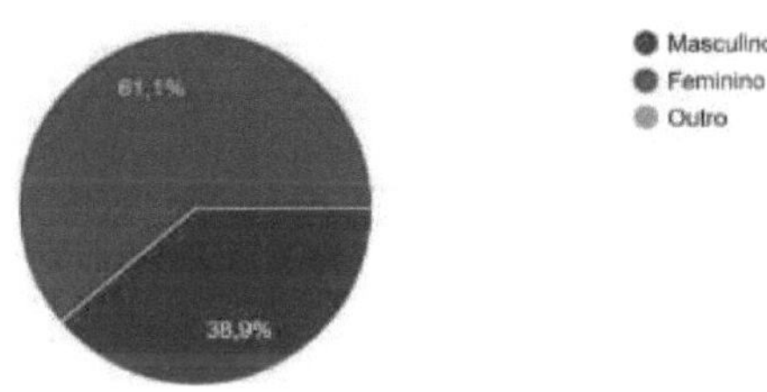

Figure 5: Population interviewed by gender
Source: Author,2022

Annex II - Population interviewed by age group

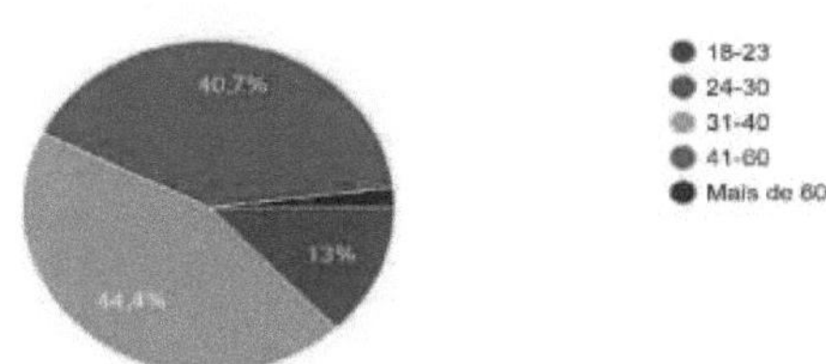

Figure 6: Population interviewed by age group
Source: Author, 2023

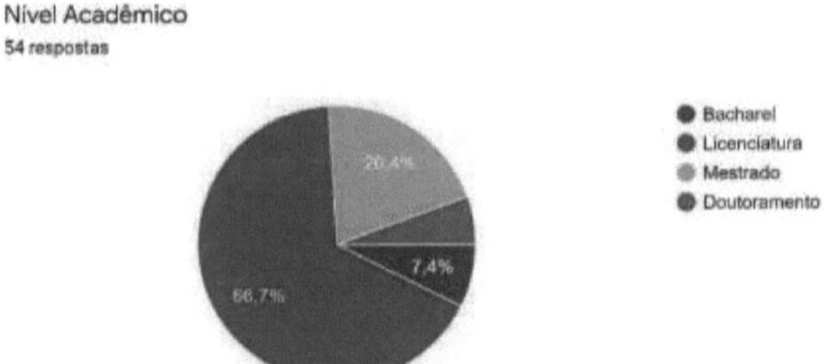

Figure 7: Population interviewed in relation to academic level
Source: Author, 2023

Annex IV - Population interviewed in relation to occupation

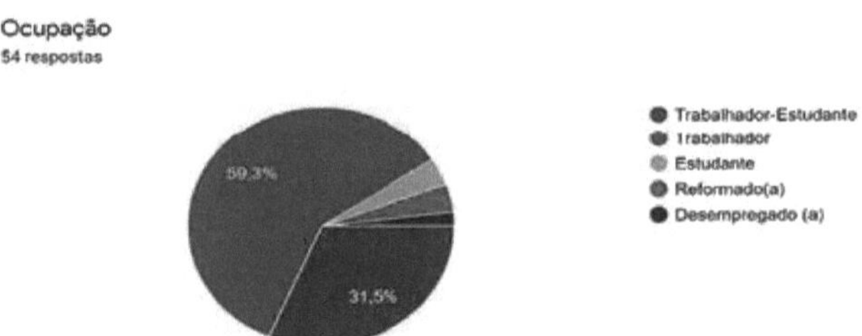

Figure 8: Population interviewed in relation to occupation Source: Author, 2023

Annex V - Would you like to take part in the online consultation?

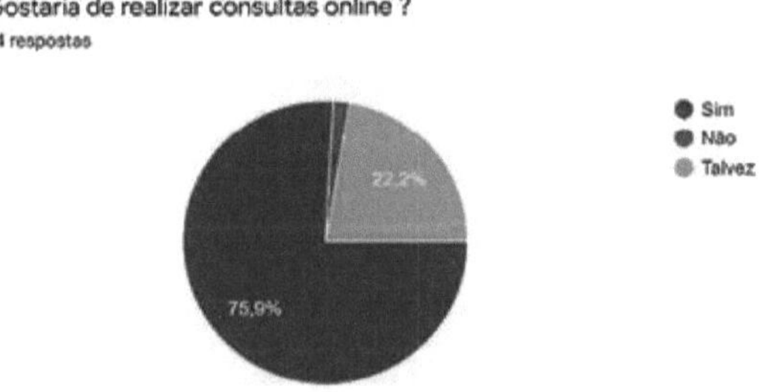

Figure 9: Would you like to make an online enquiry?

Source: Author, 2023

Annex VI - Login code

Figure 10: Login code Source: Author, 2023

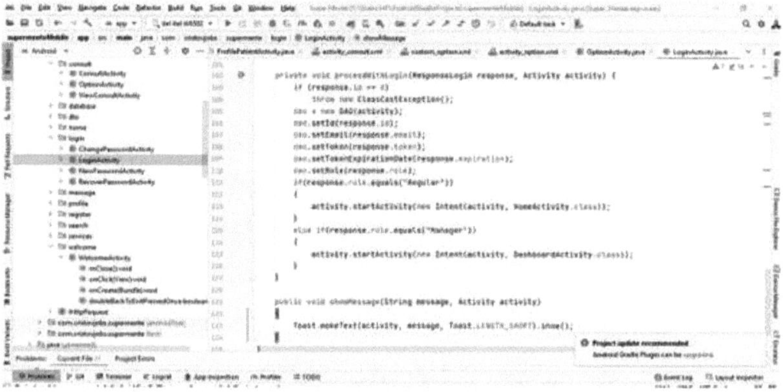

Annex VII - Password recovery code

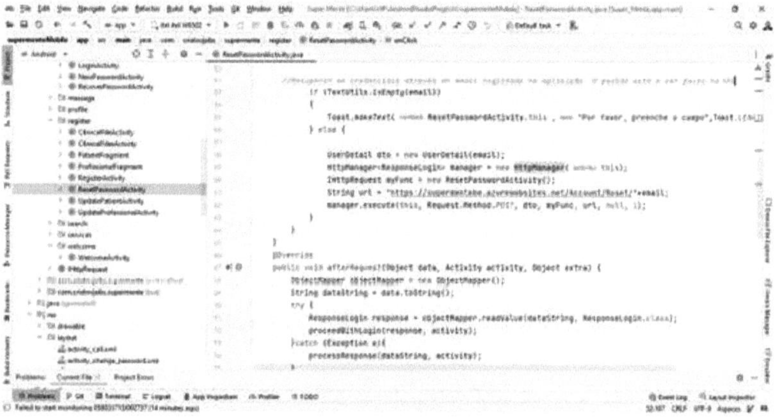

Figure 49: Password recovery code

Annex VIII - Code for registering professionals

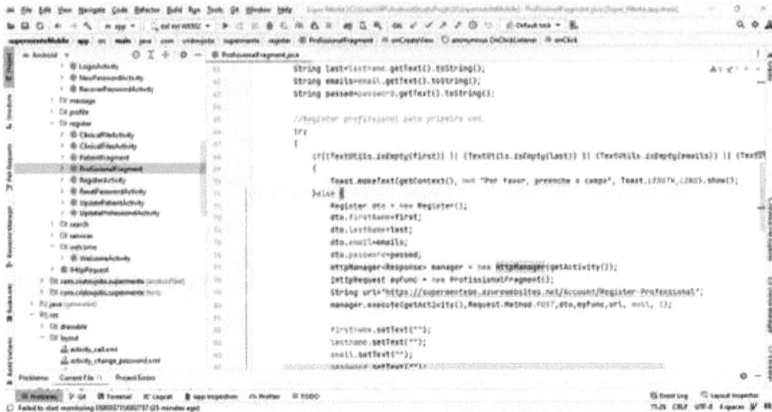

Figure 50: Code for registering a professional Source: Author, 2023

Annex IX - Code for patient registration

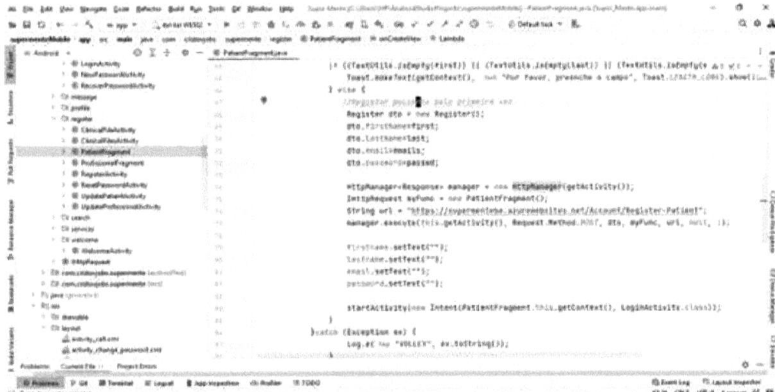

Figure 51: Code for registering a patient Source: Author, 2023

Annex X- Dashboard code

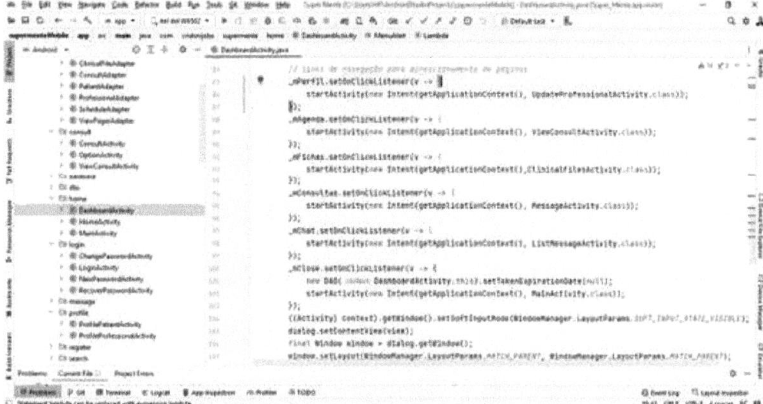

Figure 52: Code for dashboard Source: Author, 2023

Annex XI - Home page code

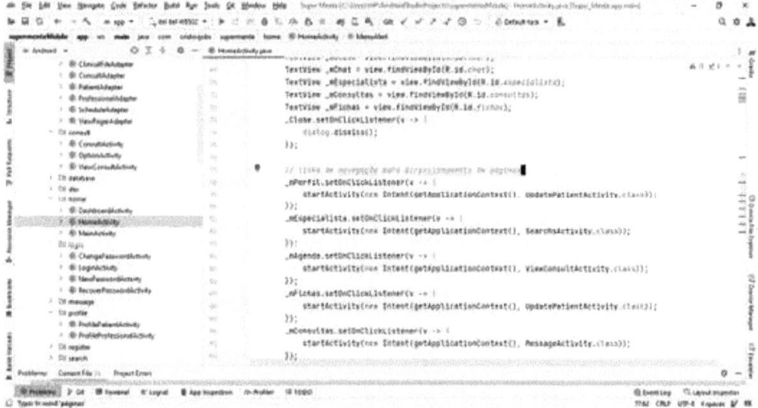

Figure 53: Home page code Source: Author, 2023

Annex XII - Patient profile code

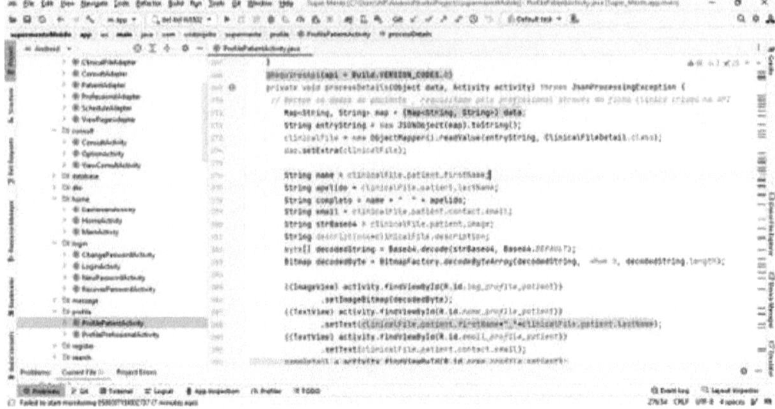

Figure 54: Patient profile code Source: Author, 2023

Annex XIII - Professional profile code

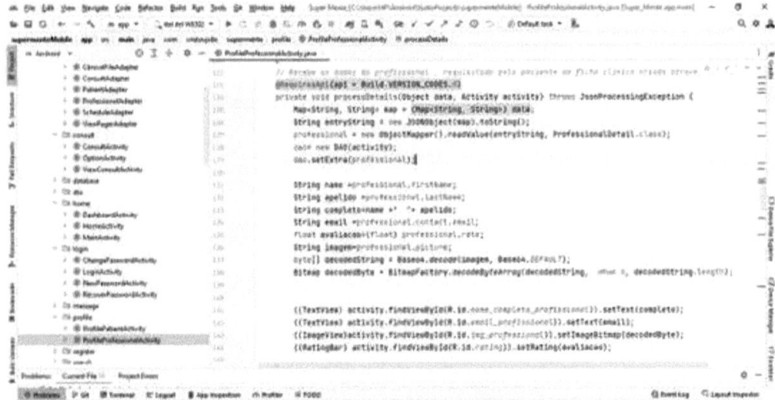

Figure 55: Professional profile code Source: Author, 2023

Annex XIV - Code for updating patient data

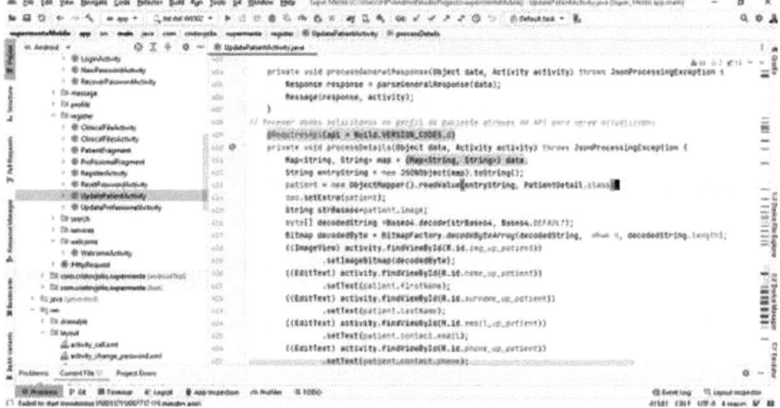

Figure 56: Code for updating patient data Source: Author, 2023

Annex XV - Code for updating the professional's data

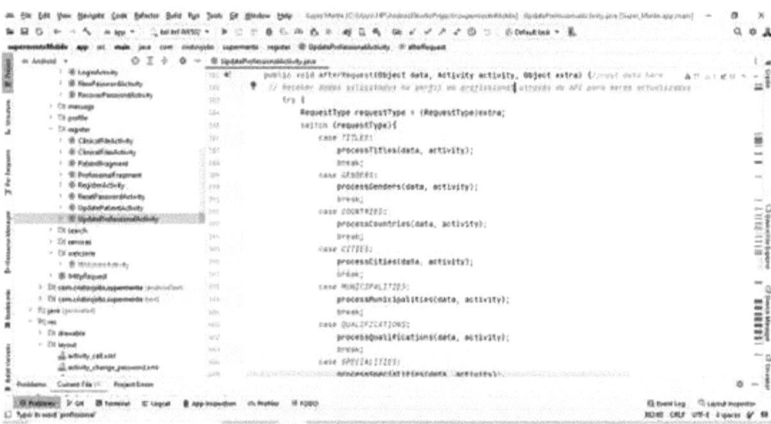

Figure 57: Code for updating the professional's data Source: Author, 2023

Annex XVI - Adapter code for the medical record

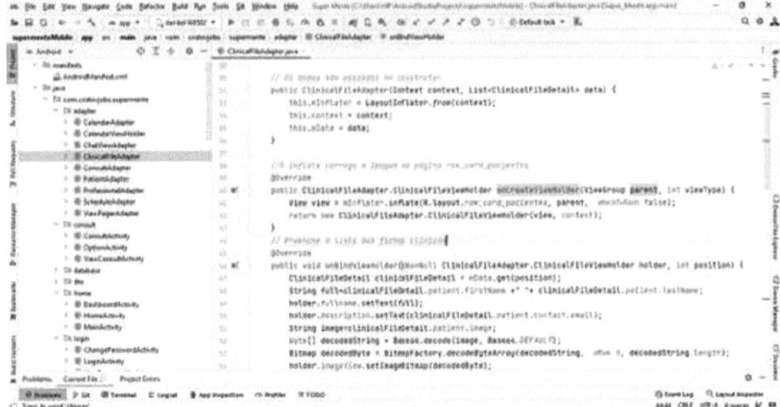

Figure 58: Adapter code for the medical record Source: Author, 2023

Annex XVII - Code to view the list of medical records

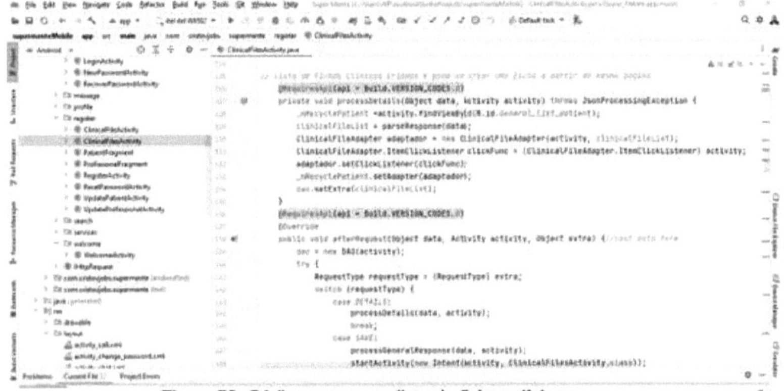

Figura 59: Código para ver a lista de fichas clínicas
Fonte: Autor,2023

Figure 59: Code to view the list of medical records Source: Author, 2023

Annex XVIII - Code for listing professionals

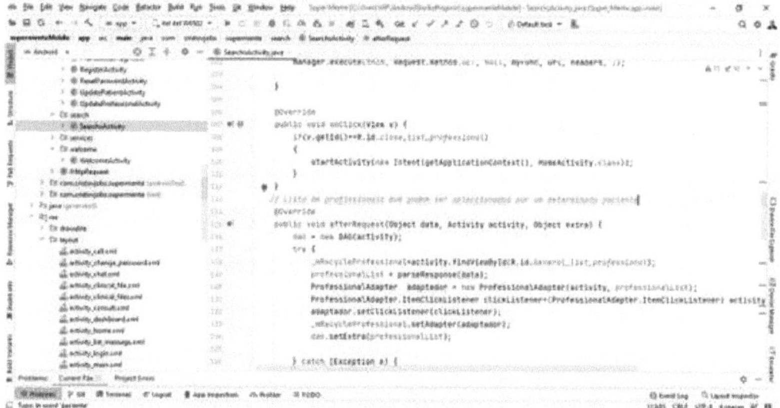

Figure 60: Code for listing professionals Source: Author, 2023

Annex XIX - Field of Study (CRER E SER Clinic)

Figure 61: Field of study - Crer e Ser clinic Source: Author, 2023

Figure 62: Field of study - Crer e Ser clinic

Printed by Books on Demand GmbH, Norderstedt / Germany